Energy, the Environment and Climate Change

Energy, the Environment and Climate Change

Peter E Hodgson
University of Oxford, UK

Imperial College Press

Published by

Imperial College Press
57 Shelton Street
Covent Garden
London WC2H 9HE

Distributed by

World Scientific Publishing Co. Pte. Ltd.
5 Toh Tuck Link, Singapore 596224
USA office: 27 Warren Street, Suite 401-402, Hackensack, NJ 07601
UK office: 57 Shelton Street, Covent Garden, London WC2H 9HE

British Library Cataloguing-in-Publication Data
A catalogue record for this book is available from the British Library.

ISBN-13 978-1-84816-415-4
ISBN-10 1-84816-415-7

Typeset by Stallion Press
Email: enquiries@stallionpress.com

Printed in Singapore.

Obituary

Peter Edward Hodgson (1928–2008)

"We have lost a great scientist, friend and person, a philosopher whose gentle ways, deep insights and infectious enthusiasm enriched our lives. We are saddened by his departure, but will retain warmth in our hearts forever as a result of having known him."

Peter E. Hodgson, Lecturer in Nuclear Physics and Fellow of Corpus Christi College, Oxford, passed away on 8 December 2008 aged 80. He was born in London in 1928 and graduated in Physics from Imperial College in 1948.

After receiving his BSc he began research as an experimentalist under the guidance of Sir George Thomson during which he was one of the first to identify the K^+ meson and its decay into three pions giving, at the time, the most accurate value of its mass. For this work Peter collected his PhD in 1951, thereafter turning to nuclear physics with H.S.W. Massey at University College London. There he studied the scattering of neutrons by alpha particles, an investigation which, when he moved to Reading, led him to explain the emission of alpha particles by heavy nuclei in nuclear emulsions bombarded with 100 MeV protons. This work raised the interest of Professor R. Peierls and Sir Denys Wilkinson who, in 1958, invited Peter to Oxford, where he become Head of the Nuclear Physics Theoretical Group and Fellow of Corpus Christi College, staying there until his retirement.

During his early years at Oxford, Peter was awarded the degree of DSc by the University of London. He also published several review papers and books, including *The Optical Model of Elastic Scattering*, which for many years became reference texts for scientists working in the field. In addition to approximately 350 original articles he wrote eleven textbooks which have been an invaluable source of inspiration to two generations of nuclear physicists.

Peter also spent much of his life devoting time to studying and promoting the impact of science on society and its moral obligation. He was an active member of the Atomic Scientists' Association serving on its Council from 1952 to 1959 and editing its journal from 1953 to 1955. In later years he became President of the Science Secretariat of Pax Romana, the bulletin of which he edited. He encouraged Catholic scientists to integrate their studies and belief and to publicise their work effectively, emphasizing the need for the Church to be thorough and professional with regard to the use of scientific advice and comment. He worked closely with the Templeton Foundation, the Newman Association and many other organisations to promote the integration of science and religion.

Only two months before his passing away, Peter wrote a letter to his friends discussing ideas for new courses on 'Physics for Philosophers', 'Philosophy of Science' and 'Effects of Science on Society' at the Gregorian University in Rome. He was writing new books: one on 'Energy, The Environment and Climate Change' and another on Galileo, which should appear soon.

> "Peter was always willing to give guidance and assistance, and lead by example. He was also the perfect gentleman."

> "He was extremely kind and caring and at the same time inspiring to a large number of young researchers. In Peter's going we have lost a precious gift of God and I have lost a great friend. He was like an older brother to me, advising, helping and inspiring me at every stage of my career. Whatever I am today Peter has made a huge contribution in all of this. My whole family is indebted to him."

Peter addressed the Vatican's 'Jubilee for Scientists' conference in May 2000 and was consultant to the Pontifical Consilium for Culture.

Peter achieved eminence in his scientific work and strove to play his part in the life of the Church to the full. All who had the great fortune of knowing him personally will sorely miss such a great scientist and a great man, husband, father and grandfather.

Peter is survived by his wife, Anne, and his four children, Louise, Mark, Dominic and Julian.

May he Rest in Peace.

(An obituary compiled, by Louise Martine [née Hodgson] from some of the many wonderful messages that Peter's family has received from colleagues across the world: Professor Ettore Gadioli, University of Milano; Professor Werner Richter, University of Stellenbosch; Professor Wasi Haider, Aligah Muslim University; Professor Anton Antonov, Institute for Nuclear Research and Nuclear Energy, Sofia.)

Acknowledgements

I would like to thank the original publishers of some of the figures in this book for kindly permitting me to reproduce them. In some cases it was not possible to contact the Editors, and so I trust that they will accept my apologies. Due acknowledgment will, of course, be given in future editions or reprints of this book should contact be made.

Preface

Mankind has lived on the earth for hundreds of thousands of years in relative harmony with their natural surroundings. The earth and its climate were unaffected by the activities of early man. Within the last two hundred years, however, this peaceful co-existence has drastically changed as a result of our scientific knowledge and its widespread technological application. New agricultural techniques have greatly increased the productivity of the land and enabled the population to rise rapidly. The industrial revolution of the nineteenth century greatly increased the living standards in many countries, but at the same time it has polluted the earth to an unprecedented degree. This pollution is changing the face of the earth and its climate at an unforeseeable rate. If it is not checked our whole civilisation is in peril.

At the basis of these changes is the demand for more and more energy to drive our industries, to heat our homes and to power our transport and communications. All known ways of generating this energy affect the earth in one way or another, by using up the energy stored over geological timescales as coal or oil and by the pollution they cause. These sources of energy will ultimately be exhausted, but if we continue to rely on them we may well cause irreversible climate change. It is therefore a matter of urgency to find safe and clean ways of generating energy. At the same time it is necessary to reduce and if possible eliminate all the other sources of pollution.

When we look to the future it is useful to distinguish between ensuring adequate energy for our needs, and the effects on the earth of the methods we use to obtain that energy. Considering the astonishing technological developments of the twentieth century and the impossibility of predicting the advances that will be made in the twenty-first century, it is unrealistic to look more than about fifty or a hundred years ahead when considering energy generation. When however we are considering the effects on the earth itself what we do now often has effects that will persist indefinitely. If the earth is polluted, it often remains polluted for a very long time. It is therefore of vital importance to ensure that current and future methods of energy generation do not irrevocably degrade the earth. To explore both these problems in more detail, the methods of generating energy by non-renewable and renewable sources are considered in Chapters 2 to 4, safety in Chapter 5 and the corresponding effects on the earth in Chapters 6 and 7. The associated political and moral problems are considered in the remaining chapters.

There is intense debate about the choice of new energy sources; should we rely on nuclear power, or could we get the energy we need from the so-called renewable sources, particularly wind and solar? As in most technological decisions, a balance has to be struck between the competing demands of cost, safety, reliability and effects on the environment. As these are incommensurable there is no easy way to reach a generally acceptable decision. It would be difficult enough to decide the optimum balance of energy sources by a dispassionate objective analysis but the whole decision process is made far worse by psychological, emotional and political forces. This makes it very likely that the wrong decisions will be taken, with disastrous effects in the future. Since countries differ greatly in their natural resources and industrial capacity no one solution is generally applicable; each county has to decide its own energy policy. Inevitably this has a great effect on international relations, particularly concerning the availability of oil during the next few decades, and of coal thereafter. In addition, the increasingly sophisticated energy technologies originate in the developed countries and are then exported worldwide. This implies continuing dependency, and with it the dangers of economic imperialism.

These decisions are not just matters of economics or politics; they raise serious moral problems. How, for example, do we decide whether to increase the level of expenditure on safety measures, or on protecting and conserving the environment, knowing that this inevitably means less money for education or the medical services? To what extent should we take account of people's emotions and fears, knowing that to a large extent they are unjustified and have been stirred up for political purposes?

The technological problems concerned with energy production are highly complex, and adequate understanding of them requires extensive scientific knowledge. It is one of the perils of democracy that vitally important decisions have to be taken by people who lack this knowledge. This problem is extremely difficult, but at least some of the worst effects can be mitigated by scientists providing whatever information they can. It is thus the responsibility of scientists to make their knowledge available by writing, lecturing and generally contributing to the public discussion of these vital issues.

This responsibility was keenly felt by those scientists who had participated in the development of the atomic bomb during the second world war. They knew that the discovery of fission and the chain reaction had irreversibly transformed the whole future prospects of the human race. On the one hand, the atomic bomb provided a weapon of unprecedented power that could, given time, be made by any medium-size industrialised country. On the other hand, nuclear power opened the way to the provision of world energy needs as the current sources, coal and oil, became exhausted. Immediately after the war, the scientists who had worked on the bomb formed organisations to inform the public of these developments; the Federation of Atomic Scientists in the USA and the Atomic Scientists' Association in the UK. They were supported by practically all the most eminent nuclear physicists, and were soon joined by scientists working in related areas. These scientists, and many others, wrote articles for magazines, organised exhibitions and gave lectures on the potentialities of the new knowledge of the atomic nucleus. The two organisations mentioned above published the Bulletin of the Atomic Scientists and the Atomic Scientists' Journal containing articles, discussions and book reviews.

Initially, these activities were welcomed by the public, and journalists wrote enthusiastic articles on the coming atomic age. Scientists, wishing to spend more time on their research and thinking that their work was done, gradually slackened their activities. They hoped that their work would be continued in a responsible way by the new generation of scientifically-trained journalists. In this they were sadly mistaken. As will be related in more detail, the public debate was soon overshadowed by political forces, and the scientists were no longer listened to. This situation persists to the present time, to the great peril of our society.

Subsequently, the Atomic Scientists' Association was wound up, and its activities transferred to the newly founded Pugwash Movement, which continues today. This provides a worldwide forum for a much wider discussion of science and public affairs. Its members now include physical and biological scientists, politicians, military men and all those concerned with international relations. Initially its main concern was to prevent the outbreak of nuclear war, and it arranged high-level meetings between Soviet and Western scientists. These were able to agree on basic scientific issues, and by communicating them to their respective Governments helped to encourage realistic policies. In the following years its activities broadened to include studies of a wide range of subjects concerned with the effects of science on human society.

I first became involved in these activities as a young graduate student, was a member of the Council of the Atomic Scientists' Association from 1951 to 1955 and edited the Atomic Scientists' Journal from 1952 to 1954. Subsequently I joined the Pugwash Movement and participated in some of its activities. This has given me some insight into the problems associated with the application of scientific ideas to the needs of society, and I am grateful to many colleagues in this work, especially to Joseph Rotblat, for their inspiration and example.

For many decades the development of nuclear power has been bitterly and successfully opposed by advocates of various renewable energy sources, particularly wind and solar. They maintain that these 'green' energy sources are less harmful to the earth and are capable of supplying all our energy needs. This claim needs to be critically examined for if it is true then it is clearly the best way forward, but if it is not then we are imperilling our future by advocating it. This debate between the proponents of the nuclear and the renewable energy sources is one of the principal themes of this book.

Many of the debates concerning the energy crisis, global warming and climate change are driven more by political convictions than by knowledge of the relevant scientific and technological facts. In many areas these facts are established beyond reasonable doubt, and in others there are still grounds for legitimate disagreement. There are disagreements about how long the coal and oil will last, and about the reality of global warming and climate change. It is therefore vital to distinguish between established facts, reasonable disagreements and pure speculation, always bearing in mind the likelihood that our judgements may be affected, whether we realise it or not, by our political and moral beliefs. Finally there are aesthetic beliefs for example about the beauty of windmills and wildernesses that cannot be resolved by any rational criteria.

Even the appointment of a Royal Commission, the time-honoured way of tackling, or quite often postponing, a decision is not always objective, as much depends of who is included in its membership. Thus the Royal Commission on 'Environmental Pollution: Energy, the Changing Climate' was chaired by an eminent biochemist and 12 of its 16 members were environmental or life scientists. They made clear their instinctive aversion to nuclear power, and this influenced many of their judgements. Nuclear power was seen only as a possibility to be accepted if all else fails. The scenarios they consider assume carbon sequestration (see Section 2.2) or final energy demands in 2050 50% lower than expected on present trends. They accepted the sensational accounts of the Chernobyl accident rather than the scientific assessment of the United Nations Commission, which found that 'apart from non-fatal and treatable cases of thyroid cancer in children, there was no evidence of other radiation related health effects in the 14 years since the accident', whereas they admitted that in the same period 'there have been some 330,000 deaths related to fossil fuel combustion in the United Kingdom alone'. They say that 'new nuclear power stations should not be built until the problem of managing nuclear wastes has been solved to the satisfaction both of the scientific community and the general public', thus ignoring the view of the OECD Nuclear Energy Agency that 'there is broad scientific consensus that the disposal of high-level long-lived radioactive wastes in deep geologic formations is an appropriate and safe means of isolating it from the biosphere for very long timescales' (see Section 4.6). If the Commission had a different membership the conclusions would probably have been quite different (Nuclear Issues 22, September 2000).

It is important to recognise that it is very easy by selective quotation to make a strong case either for or against subjects like the desirability of nuclear power, the existence of global warming or the so-called population explosion. These are complex issues and it is impossible to reach a sound conclusion without extensive knowledge. Lacking this, the next best thing is to follow the conclusions of panels of experts convened by scientific societies such as the Royal Society and the National Academy of Sciences. Otherwise it is easy to be misled by the many accounts that are published by scientists who are indeed very distinguished, but generally not in the speciality they are writing about. Further investigation may show that they are influenced by political or financial considerations. There is, for example, the well-known case of a distinguished solid state physicist who distributed large sums from the tobacco industry to researchers who found no harmful effects of smoking. Extreme care is necessary to obtain an impartial and balanced view.

It is quite extraordinary that many excellent books on the energy crisis, global warming and climate change, such as those by Gore and Maslin, make only the briefest references to nuclear power, brushing it aside with a few critical remarks about nuclear accidents and the disposal of nuclear waste. Conferences arranged by the British Government on the best ways to tackle global warming have many sessions devoted to wind, solar and wave power, with strong recommendations to improve energy efficiency, but fail even to mention nuclear power. The mass media show a similar bias, giving front-page publicity to the most minor nuclear accidents, while barely mentioning major disasters claiming hundreds of lives in dam bursts, oil rig fires and collapse of coal mine tunnels. Since nuclear power is a major source and is non-polluting, it would seem that it is necessary to consider their arguments against it, instead of ignoring it entirely. There seems to be a widespread and deep-seated psychological aversion to nuclear power. To redress the balance this book takes nuclear power seriously and discusses it in some detail. The emphasis is on the scientific and technological aspects, while some of the economic aspects concerning the buying and selling of companies building reactors receive only a brief mention.

One might expect there to be strong correlation between the scientific and technological feasibility, cost, reliability and safety of an energy source and its public approval, together with Governmental support for its development. This is far from the case. Political and psychological pressures are often far more influential than proven scientific data. It is possible to ignore reality for a time, but the longer this is done the more severe the ultimate reckoning. As Feynman remarked, Nature cannot be fooled.

These problems are of serious concern to the more well-developed countries, but they are a matter of life and death for the poorer ones. Already climate change is believed to be causing widespread drought, and with it famine and disease. Most of these countries lack both the will and the means to improve their situation, so it can be maintained that it is the duty of the developed countries to do all they can to improve the living standards of the people in the poorer ones. There are many serious difficulties in achieving this, but they need to be urgently tackled.

In all discussions related to energy it is essential to express the quantities discussed numerically as accurately as practicable. The capacities reliabilities and costs of the various energy sources can be expressed fairly accurately, and also to some extent their safety, expressed as the numbers of persons killed or injured. It is more difficult to express the effects on the environment, as these involve aesthetic criteria about which legitimate differences exist. The numbers I quote are the best that I could find, and I will be grateful to receive better ones. It is unfortunately inevitable that in the present situation where events are changing rapidly that many of the numbers are outdated, but nevertheless they serve to indicate general trends.

My main concern throughout is to draw attention to some of the most pressing problems of the present time, to stimulate discussion and to emphasise the moral aspects. The primary responsibility of scientists is to explain the scientific facts and their technological implications. In some cases, once the facts are known, the way forward is obvious, in others any attempt to provide an answer would be premature. Scientists as such have no responsibility to decide moral questions; that is the responsibility of the whole of society, including the scientists in their capacities as citizens. In order to reach sound decisions on moral questions expert and authoritative guidance is needed, and this should be provided by moral theologians and Church leaders. In order to give this guidance, it is essential that they are fully aware of the relevant scientific and technological facts This seems so obvious that it might be considered to be hardly worth saying, but what is nearly always lacking is a full appreciation of what this really means. The necessary knowledge cannot be obtained easily; it requires years of study and research. The lack of understanding of this basic point has rendered worse than useless most of the statements by doubtless well-meaning Church leaders.

When I first became interested in these problems the main concern was to ensure that there is sufficient energy to maintain our standard of living and to raise that of the people in poorer countries to the same level.

It gradually become clear that there are sufficient energy resources to do this, mainly in the form of coal and uranium, but then the pollution caused by fossil fuels became the subject of concern, together with the likelihood of gradual and perhaps catastrophic climate change. To avoid this outcome very drastic action has to be taken and the main concern now is whether it is politically or psychologically possible to make the necessary changes to our styles of living in time to avert disaster.

During the last decade scientists have made increasingly accurate forecasts of the dangers threatening the world and the actions that must be taken to avert them. More and more people are becoming convinced that these actions are necessary, but virtually nothing has been done. Governments have indeed set up committees to examine these problems and make recommendations, but they are subsequently ignored if it seems politically expedient to do so. We are heading into disaster with our eyes open.

In writing this book I have used the information in many books and articles particularly those of Lomborg, MacKay and Maslin, and also the periodicals Nuclear Issues and Speakers' Corner, and the SONE (Supporters of Nuclear Energy) news sheets.

I gratefully acknowledge the help of many colleagues and especially Dr. D.A. Hodgson.

P.E. Hodgson
2008

General References

Gore, Al (2007) *Earth in the Balance* (London; Earthscan).

Hodgson, Peter Edward (1999) *Nuclear Power, Energy and the Environment* (London: Imperial College Press).

Lomborg, Bjorn (2004) *The Skeptical Environmentalist* (Cambridge: Cambridge University Press).

MacKay, David J.C. (2008) *Sustainable Energy; Without the Hot Air*.

Maslin, Mark (2004) *Global Warming* (Oxford: Oxford University Press).

Contents

Chapter 1

The Energy Crisis

1.1. Introduction

Our civilisation and our standard of living depend on an adequate supply of energy. We need energy to light and heat our homes, to cook our food, to drive our transport and power our communications and to provide the motive force that drives the factories. Without energy all this would be impossible on the scale needed, and our civilisation would soon collapse into barbarism. Our dependence on energy is strikingly illustrated by the connection between average life expectancy and energy consumption. People in the poorer countries, especially in Africa and Asia, have an average energy consumption between 0.01 and 0.1 tons of coal equivalent per person per year and have an average life expectancy of between 35 and 45 years. At the other end of the scale, people in the rich well-developed countries in Europe, North America and Japan use between five and ten tons of coal equivalent per person per year and have an average life expectancy between 70 and 75 years. This difference is a measure of the energy that is needed to bring the standard of living of all people to the level now enjoyed by the most favoured ones.

Over the centuries this energy has been obtained in many ways. In ancient times wood was the main fuel, and it provided heat for cooking and warmth. It was often used more rapidly than it was replaced by new growth, and the forests of countries surrounding the Mediterranean were gradually destroyed, followed by the forests of central Europe. In many countries even today wood is the main fuel, but housewives have to walk increasing distances to gather the wood they need. Other energy sources are crop residues and dried animal dung. Ideally these organic residues should be returned to the soil, so burning them gradually reduces its fertility. These are still the main sources of fuel for some

1

two billion people in the developing countries. They provide the energy equivalent of about a billion tonnes of oil each year, about the same as the energy provided by coal in Europe and the USA combined.

The increasing scarcity of wood stimulated searches for alternative energy sources, and soon coal was found, first near the surface and later underground. It has a higher calorific value than wood and can be transported rather more easily. Soon it became the main energy source in many developed countries and provided the power for the industrial revolution, especially in places where iron ore was also available.

During the nineteenth century oil was found, first in the USA and then in many other countries. It has many advantages over coal: it can easily be transported over large distances by pipelines and tankers, and is the basis of the petrochemical industry. During the twentieth century it gradually displaced coal as the favoured energy source. Natural gas was often found in association with oil, and provided a convenient source of lighting and heating.

The nineteenth century also saw the rapid development of the electrical industry for communication, heating and power. Electricity has the advantage of being very easily transported from the generating station to where it is needed. It soon displaced gas as a source of light and became a convenient power source for factories. Electricity is practicable for suburban trains, but long distance trains and ships, which used to be driven by coal, are now mainly driven by oil.

Electricity is generated by turbines driven by steam produced by burning coal or oil. The turbine can also be driven by water, and indeed water wheels have been used since ancient times to rotate the millstones to grind corn. Hydroelectric power is thus another source of electricity.

During the twentieth century the world's economy and population increased more rapidly than ever before and the total energy consumption rose even more rapidly. World population is doubling on the average every 35 years; the rate of increase varies greatly from country to country, it is greatest in Africa and Latin America and almost stationary in more developed areas such as Europe and North America. Together with the increase in the standard of living, this results in the world energy consumption doubling every fourteen years. This is not a measure of the real energy needs, and many billions of people still lack the energy for even some of the basic necessities of life. At present people in the less developed countries are forced to try to survive on a small fraction of the energy used by people in the developed world. The amount of energy needed to raise the standard of living of the people in the poorer countries to that in the developed ones can be estimated from the United Nations Human Development Index. This shows that an acceptable standard of living requires about five thousand kWh per year, or 200,000 megajoules (MJ) per person. Assuming that the world population will rise to eight billion this gives an energy requirement of about 1.6×10 (15) MJ. The present population of six billion uses about 0.41×10 (15) MJ. Thus world energy production will have to be increased at least fourfold to bring the standard of living of people in the developing countries up to that in the developed ones (Fanchi 2006). This estimate does not take account of the likelihood that most of the increase in population will take place in the poorer countries. It has been estimated that world energy production will grow from 4.43×10 (14) MJ in 2003 to about 7.6×10 (14) MJ by 2030, which is completely inadequate. Another estimate is an increase from 9.3 billion toe (tonnes of oil equivalent) in 2003 to 15.4 in 2020, with 90% of the increase in the developing countries.

The world consumption of energy increased by 4.3% from 2003 to 2004, and this trend is likely to continue. At present the relative amounts due to the various sources are: oil 36.8%; gas 23.7%; coal 27.2%; nuclear 6.1% and hydro 6.2%. Of these coal is the fastest growing and also the most polluting. The main sources of energy at the present time, coal, oil and gas, are limited and indeed are fast being exhausted. Studies of the available resources indicate that in the foreseeable future they will become increasingly difficult to extract in the quantities needed. Furthermore, they are seriously polluting and are already causing great damage to the earth and its atmosphere. Eventually we shall have to learn how to do without them, and the sooner we do so the better.

The generation of electricity is expected to show a similar rise, increasing from 13,290 billion kWh in 2001 to 23,702 in 2025. In the developing countries the rate of increase is 3.5% per year, compared with 2.3% per year for all countries. The proportion of electricity generated by natural gas increased in the same period from 18% to 25%.

What can replace them? There are many possibilities, and they have to be assessed considering their capacity, cost, reliability, safety and effects on the environment. This is done in the following sections. Whenever possible, these assessments must be made numerically because this is the only way to make objective comparisons. As Lord Kelvin once remarked, *"unless you can measure what you are talking about, and express it in numbers, your knowledge is of a meagre and unsatisfactory kind"*. When evaluating the characteristics of the production of energy by different sources these numbers are often subject to many uncertainties, so it is also important to estimate the range of these uncertainties. Approximate numbers are better than no numbers at all. The existence of numerical data is not however a guarantee of its correctness. Many numbers widely quoted are simply wrong, and sometimes numbers that are correct are extrapolated far beyond their range of validity and used as the basis of inaccurate generalised statements. It is thus not enough to collect a few isolated examples and assume that they are representative of global trends. There is no way of avoiding the laborious task of collecting fully representative statistics and recognising that they can only be used to reach valid conclusions for the time and area actually studied.

It is useful to collect and compare numerical estimates of the same quantity obtained by different investigators as, for example is done in Table 3.3 for the costs of electricity from various sources. Physicists know very well how difficult it is to obtain a reliable measurement of some physical quantity such as the charge on the electron, when no one has any motive for preferring one result rather than another. It is quite different when comparing energy sources, because there may be strong political pressure to reach a favoured result. Furthermore, there is an inherent difference between the degree of accuracy that can be reached when measuring physical quantities, and that attainable in a social context. Thus the accuracy of determination of the charge on the electron can presumably be increased without limit, whereas there is an inherent limit to the values of social parameters. Thus we can collect statistics relating to a specific stretch of time at a certain place. We cannot improve the statistics by extending the period of time because the method of energy production may itself have changed; we cannot be sure that the quantity we are measuring is independent of time and place and this inevitably affects the accuracy with which it can be determined. The figures given here are the best I could find, but certainly in many cases better figures will eventually become available.

Energy sources are often divided into renewable and non-renewable. A renewable source is one that is not used up, but what we really want is one that is always available. This is satisfied by a source such as water, the fuel for fusion reactors, that is so plentiful compared with our needs that it will never be exhausted, and also to a lesser extent uranium. Coal and oil are non-renewable because there are limited amounts present in the earth. However the reserves of oil and coal are not like the gold in the Bank of England, where the number of bars can be counted. The amount we can extract depends on the price we are prepared to pay for it. Oilfields, for example, have very different extraction costs. In the Middle East, oil gushes out freely and is cheap and readily available. It is much more expensive to extract it from the North Sea, as oil rigs have to be built in deep water. This consideration applies even more strongly to minerals such as those containing uranium. Rich ores are relatively rare, while poorer ones are very widespread. It is even possible to extract uranium from sea water.

It might be thought that the lifetimes of availability of an energy source can be obtained by dividing the estimated reserves by the yearly consumption. This gives about 42 years for oil, 65 for natural gas and 217 for coal. Another way of expressing this is to divide the total recoverable resource of 1260 TWy (terawatts per year) by the world consumption, which amounted to 14.1 TWy in 2003. This gives about 90 years. By 2030 the world consumption is expected to be about twice this, halving the expected lifetime (Avery 2007, pp. 107, 113). However the situation is not as bad as this because there are several competing effects that occur when any raw material becomes more difficult to obtain. Initially the price rises, providing an incentive to reduce consumption and to look for alternatives sources. Improved technology and the higher price make it economic to bring into operation sources previously considered to be exhausted. The Stone Age ended because bronze and iron were discovered, not because the supply of flints ran out. This process continues: steel and other metals are replaced by plastics and composites. In other cases the incentive for change comes from other motives, such as the replacement of domestic coal fires by electric heaters and much correspondence by e-mail. The net result of all this is that, contrary to what might have been expected, it is found for coal and oil that the ratio of reserves to annual consumption and the cost both remain remarkably steady as a function of time (Lomborg 2001/2004), as shown in Figures 1.1 and 1.2. Of course this cannot go on for ever, but it is difficult to estimate the lifetimes, although prices are certainly expected to rise.

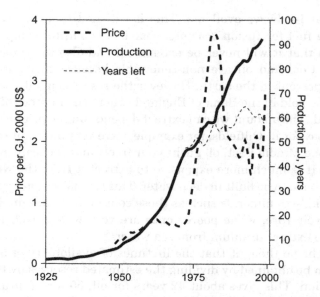

Figure 1.1. World gas production, price and years of consumption. Production in exajoule, 1925–1999, price in 2000 US$ per gigajoule, 1949–2000, and years of consumption, 1975–1999 (Lomborg 2004).

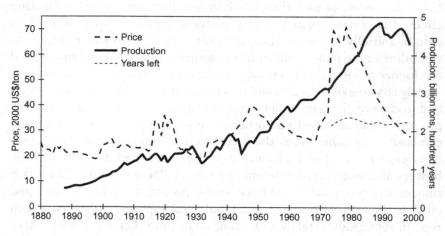

Figure 1.2. World coal production, price and years of consumption. Production in billion tons, 1888–1999, price in 2000 US$ per ton, 1880–1999, and years of consumption, 1975–1999 in hundreds of years (right axis) (Lomborg 2004).

It is also difficult to obtain accurate estimates of the cost of energy from various sources. It is relatively easy to determine the cost of construction of a power station, but not to foresee its working life. A power station may last fifty years or more, and it is not easy to determine the lifetime in advance. During that time the value of the currency is likely to change substantially which makes it difficult to calculate the true cost of the electricity generated.

For most industrial and domestic uses, particularly when electricity is being used, it is essential that the supply is reliable. The use of computers is spreading, and an interruption of electrical power immediately brings many vital activities to a standstill. Industrial processes controlled by computers come to a stop, electronic communications are cut, and supermarkets have to close. The cost of such interruptions, even if quite short, can be very high. If the breakdown is prolonged, households relying on electricity are in severe difficulties as heaters, lights and refrigerators fail. We are so used to relying on our power supplies that we are thrown into disarray when they cease. In New York a few years ago, a power station was overloaded and so automatically cut out, throwing more load on another power station, which also cut out and so on until the whole electrical grid was down. The system was so complicated that it took several days before power was restored. Such events are very rare, but power breakdowns are everyday occurrences in many countries, so that it is not possible to carry out lengthy computer calculations or delicate manufacturing processes.

It is neither possible nor necessary for every power station to be able to operate indefinitely, but they should at least normally operate for several months at a time, and this is the case for coal, gas, nuclear and hydroelectric power stations. Their electrical output is fed into a national grid, so if one power station breaks down the load can be carried by the others without interrupting the supply. From time to time power stations must be shut down for routine maintenance, and the grid allows this to be done without difficulty. The demand for electricity fluctuates, but usually in a predictable way. The greatest demand is in the winter, so most maintenance shutdowns are scheduled for the summer. The greatest demand comes during sharp cold spells, so the maximum number of power stations are made available at times when this may happen.

There are some applications where the reliability of the energy source is not important. For example, a farmer may need to have water available for irrigation, so he installs a storage tank and uses a windmill to pump water from below ground to fill it. Providing the tank is large enough, and the wind blows now and then, there is always enough water when needed, and it is topped up whenever the wind is blowing. In this way the unreliability of the energy source is converted to reliability of the system. This is only possible when the amount of energy to be stored is relatively small, as it is uneconomic and often impossible to store very large amounts of energy.

Another requirement is that the energy is available in concentrated form, that is at a higher temperature than the surroundings. Every room contains a large amount of energy, but it is useless for boiling water in a kettle since the availability of energy depends on the temperature difference between source and surroundings. Energy exists in many form, kinetic, potential, electrical and chemical, and since conversion from one form to another always takes energy it should be in the form needed whenever possible. As MacKay (2008) remarks, 'you can't power a TV with cat food, nor can you feed a cat with energy from wind turbines'.

Safety is an essential requirement for energy generation. Perfect safety is impossible, so it is necessary to ensure that any source is a safe as reasonably possible. Increasing safety is costly, so a balance has to be struck between conflicting demands. The same applies to the requirement that adverse environmental effects should be minimised.

Energy is very often wasted, and it is frequently urged that the energy crisis can be solved by increasing the efficiency of energy use and eliminating waste. This is discussed in the next section and in the following sections the possible alternative energy sources are assessed according to the five criteria proposed above.

1.2. Energy Conservation

It is frequently argued that we could solve the energy crisis simply by using energy more efficiently. At present we are very wasteful. We leave lights on in empty rooms, heat parts of our houses that are not in use, and allow the rooms in use to be too hot, so that most of the heat escapes through the windows and walls. In many warm countries air conditioners are over-used in summer so that the rooms become too cold. Much larger amounts of energy are wasted by inefficient processes in factories. Huge amounts of energy are spent on unnecessary journeys and leisure activities. Advertising and high-pressure salesmanship encourages people to buy things that they really do not need, and credit schemes make it easy to do so. Changes of fashion in clothing and house-styles mean that perfectly useable items are thrown away. Many machines are not designed to last as long as possible; on the contrary, manufacturers ensure by built-in obsolescence that they will have to be frequently replaced. The aim is to increase their production and maximise their profits at the expense of the consumer, and in the process much energy is wasted. It is very often easier and cheaper to throw away a defective machine and buy another rather than have it repaired. Indeed it is now almost impossible to find people willing to repair anything.

All this is happening at a time when billions of people lack the energy needed to provide the bare necessities of life. It is immoral to allow this to continue. In any case in the long run it cannot continue as the resources of the earth are limited and thus unable to sustain indefinite growth.

It is very easy, and indeed it is a moral duty, to save energy by quite simple measures. We can switch off lights not in use, and install thermostats to keep the temperatures of the rooms at a moderate level. Fitting double glazing, lagging pipes and insulating walls and roofs also cuts down the energy loss. We can avoid unnecessary journeys, use public transport whenever possible, walk instead of driving, and avoid leisure activities that waste energy. Houses can be built to save energy and Industrial processes can be designed in a more efficient way. It has been estimated that if such measures are introduced the energy use can be reduced by a very substantial factor (Von Weizacker *et al.* 1996).

Many of these energy-saving measures can be taken immediately, but others take time and may themselves be energy-consuming. Thermostats and lagging, for example, have to be manufactured, and this costs energy. There is a time-lag before there is a net saving of energy. It is much easier to design and build an energy-saving house than to convert an existing building.

It is easy to urge that energy be saved, but far more difficult to convince people to take the necessary action. If energy is cheap, people just cannot be bothered to take energy-saving measures. Furthermore, many of the energy-saving measures themselves cost money, and even if there is a long-term gain this is a strong disincentive. It is easy to reduce energy consumption by increasing the price but this can provoke a violent political reaction, as happened recently in the UK when the price of petrol was raised. Furthermore, a rise in energy price hits the poor harder than the rich, and can seriously affect their health. Many poor people suffer from malnutrition and even die from hypothermia in the winter because they cannot afford food and fuel. Care must be taken to avoid this by measures such as reducing the price of electricity for the first few units every month and then increasing the price for higher levels of consumption.

There are also larger changes concerning the management of the economy and changes in the structure of society that could effect large energy savings and save scarce resources. Transport is a large consumer of energy in the form of petroleum, which is polluting as well as running out. It is thus very desirable to switch to other means of propulsion such as electricity. Already there are battery-driven cars and improved designs are being developed. These are designed so that the batteries can be recharged when the car is not in use. At present they are more costly than petrol-driven cars, so perhaps government intervention in the form of subsidies could encourage the widespread use of such vehicles. To avoid pollution the electricity must be generated by a non-polluting source, of which nuclear is the only practical way at the present time. It is also possible that some time in the future transport could be driven by hydrogen, as discussed in Section 3.10

Further energy savings can be obtained by encouraging the use of more energy-efficient means of transport such as trains and buses instead of cars and lorries for passengers and for freight. To illustrate this, MacKay (2008) gives the following transport efficiencies in Japan (1999) in kWh per 100 passenger-km: car 68, bus 19, rail 6, air 51, and sea 57.

As the cost of travel increases, mainly due to the rise in oil prices, it will become more difficult for people to afford to travel long distances to work, or even to have a car for themselves. This in turn will make living outside the towns and cities less attractive. This may be partly offset by the use of modern technology to make it possible for many people to work at home, at least for some of the time. This implies large changes in the structure of society, and it would be wise to bear this in mind when planning new housing developments. Another consideration is that the predicted rise in sea level makes it undesirable to build houses on low-lying land liable to flooding.

It is also necessary to ensure that energy-saving measures are not counterproductive. A consideration that is often overlooked is that using more efficient manufacturing processes can reduce the cost of production and, to gain an advantage over their rivals, manufacturers can then reduce the price of their goods and hence increase the sales. The final result is that more energy is used than before. This was already recognised by the Victorian economist Stanley Jevons who wrote: 'It is wholly a confusion of ideas to suggest that the economical use of fuel is equivalent to a diminished consumption. The very contrary is the truth. It is the very economy of the use which leads to its extensive consumption. It has been so in the past and will be so in the future.'

Many of these energy-saving measures are being implemented but nevertheless the increasing world population and the average rise in living standards are causing the world consumption of energy to rise linearly with time at the rate in 2004 of about 4.3% per year. Thus while energy conservation is extremely important and should be encouraged by all practicable means it will not by itself solve the energy crisis. It is therefore still necessary to examine all possible energy sources to see which of them can best provide our energy needs.

Quite often there is more energy available than we need at the time, so many of our energy problems would be solved if energy could easily be stored on a large scale. It has already been mentioned in connection with storing water on a farm using a windmill that inevitably operates discontinuously. When the reservoir is full and the wind still blows that energy is wasted. This does not matter because relatively small amount of energy are involved. The situation is quite different for large amounts of energy. It is then necessary to have enough power stations to meet periods of peak demand, with extra allowance for unexpected failures. At other times there is a large excess capacity, and if the extra energy generated in these periods could be stored then fewer power stations would be required. Unfortunately this is not possible because electrical energy must be used as soon as it is generated and it cannot be stored economically. Batteries are unsuitable and far too expensive for this purpose. A practicable possibility is to use excess energy generation to pump water from a lower to a higher level and then to let it flow down through turbines when demand requires. This can be done in places where there are two large lakes not far apart at very different altitudes, but there are very few such places and in addition the process itself wastes about 30% of the energy. Another possibility is to use the excess electricity to decompose water into hydrogen and oxygen and store the liquefied gases. These can be allowed to recombine when the stored energy is needed. This process is expensive and at present more so than the cost of building extra power stations.

Other possible energy storage methods are flywheels, supercapacitors and superconducting coils (Swamp 2007). One of the problems in supplying power for spacecraft is that batteries last only a few years, so it is proposed to replace them by flywheels. It is also planned to develop a 20 MW power plant to be connected to the national electricity grid. This could store energy and release it very rapidly, in about 4 seconds, when there is a surge in demand. Another application is to store the energy of heavy vehicles such as buses and lorries when they brake, and then release it when they start to move again. This could halve the fuel used by vehicles that frequently stop and start. However the cost of this device is about $50,000, three times that of a fuel cell. Supercapacitors are already used in laptops and other electronic devices. For most applications, however, supercapacitors and superconducting coils are far too expensive.

1.3. World Energy Consumption

Before considering each energy source in detail it is useful to compare their contributions to world energy consumption. This serves to keep their relative importance in perspective. Figures 1.3 and 1.4 and Table 1.1 provide some figures for past consumption and estimates for the future. The four main producers are coal, oil, natural gas and nuclear, with smaller contributions from other sources. Hydro is the next in importance but as it is limited by the number of suitable rivers its contribution remains almost constant and its relative contribution decreases. The contribution of the remaining sources is rather small. The 'traditional' renewable energy sources such as wood, straw and dung (biomass) amounted to 0.9 GTOE in 1990. Modern biomass is growing special crops such as willow for subsequent burning.

World energy consumption increased by 4.3% in 2004, with some countries increasing faster than others. Thus Chinese consumption increased by 15% in one year (Nuclear Issues, June 2005). Consumption is expected to double by 2050. China imported about 40% of its oil consumption of 250M tonnes and this is expected to increase to 60%

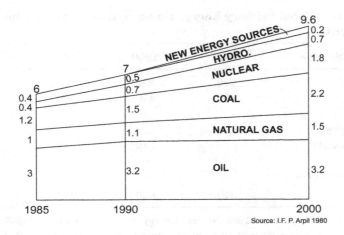

Figure 1.3. Energy Consumption Outlook in the Western World (GtOE) (World Energy Needs and Resources).

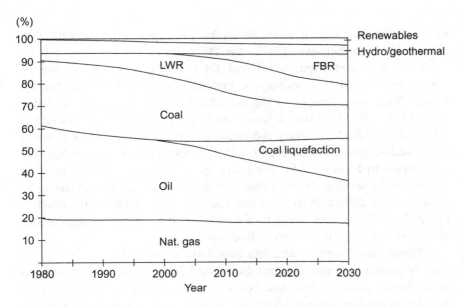

Figure 1.4. Shares of primary energy sources in the global energy balance (World Energy Needs Resources).

Table 1.1. Global Primary Energy Consumption (in million tons of oil equivalent)

Energy Source	1860	1900	1950	2000
Wood etc	270	330	470	~1000
Coal	100	470	1300	2220
Oil		20	470	3400
Natural gas			170	2020
Hydro-electric		10	120	230
Nuclear				630
Other renewables				~200
Total	370	830	2530	~9700

Sources: For 1860, 1900 and 1950: Nuclear Energy in Industry (Crowther 1957); figures converted from coal equivalent to oil-equivalent energy by dividing by 1.5. For 2000: Statistical Review of World Energy (1999 BP Amoco), trended up to 2000; except wood etc., from Rural Energy and Development (1996 World Bank). For primary energy, BP assumes that one tonne of oil produces 4000 kWh in a modern power station.

by 2020 (Nuclear Issues, June 2005). An additional 2 million cars were sold in 2003, an increase of 80% over 2002.

Comparison of Figures 1.3 and 1.4 with Table 1.1 show general agreement for past consumption, but substantial differences for the future. Thus for example, in Figure 1.3 nuclear in 2000 is estimated in 1980 to be 1.8, while the estimate in Table 1.1 from data obtained in 1999 is only 0.63. This sharp reduction in the nuclear estimate is due to the reaction against nuclear power that is discussed in Section 8.5. The projections in Figure 1.2 for the years up to 2030 are even less reliable. For example, the fast breeder reactors (FBR) are projected to rise quite sharply after 2000 and the produce more than the light water reactors (LBR) by 2030. In fact, fast breeder reactors have not been deployed at all, for the reasons discussed in Section 4.7

These uncertainties are characteristic of all estimates of future energy production and demand. Some variability is to be expected due to different availabilities and costs of raw materials, different legal requirements concerning the siting, building and running of power stations and in addition large variations may occur due to political pressures and unexpected events. It is also not unknown for figures to be carefully selected according to different criteria in order to produce a result that is already desired on political grounds. It is therefore important to collect statistics from various sources whenever possible and analyse them in detail before reaching a final estimate.

The rates of energy consumption vary from one country to another, depending on the wealth of the country, its Government and its natural resources. The energy use in Britain, Switzerland, India and the USA is given by Ramage (1997).

It is instructive to compare these figures with those for the sources of electricity in France, one of the countries with the highest nuclear component. In 2006 the production was 549 TWh, consisting of 429 (78%) nuclear; 61 (11%) hydro; 57 thermal (10%) and 2.2 Wind (0.4%) (Nuclear Issues 30, March 2008).

1.4. Wood

As already mentioned, wood is still extensively used as a major source of fuel in poorer countries. If more is used than is replaced by additional growth this can lead to desertification. It is therefore desirable to replace wood by more efficient and less damaging fuels, and certainly the use of wood cannot hope to solve the energy crisis.

References

Avery, John Scales (2007) *Energy, Resources and the Long-Term Future* (London: World Scientific).

Barnham, K.W., Mazzer, M., and Clive, B. (2006) Resolving the energy crisis — nuclear or photovoltaics, *Nature (Materials)* **5**, 161.

Hafele, Wolf Ed (1981) *Energy in a Finite World: A Global Systems Analysis* (Ballinger Publishing Company).

Hodgson, Peter Edward (1999) *Nuclear Power, Energy and the Environment* (London: Imperial College Press).

Lomborg, Bjorn (2004) *The Skeptical Environmentalist* (Cambridge: Cambridge University Press).

MacKay, David J.C. (2008) *Sustainable Energy: Without the Hot Air*.

Ramage, Janet (1997) *Energy: A Guidebook* (Oxford: Oxford University Press).

Swamp, Bob (2007) Energy storage takes off. *Physics Word*, July, p. 42.

Von Weizacker, E., Lovins, A.B., and Lovins, L.H. (1996) *Factor of Four: Doubling Wealth-Halving Resources: The New Report to the Club of Rome* (Earthscan Pulications Ltd.).

Chapter 2

Non-Renewable Energy Sources

2.1. Introduction

At present most of the energy used is generated by non-renewable sources such as coal and oil. This immediately raises the problem of what to do when they become exhausted. As it is unrealistic from the point of view of energy generation to look more than about a hundred years ahead, it is important to distinguish between sources that will last more than a hundred years and those that will not. As we shall see, oil and natural gas will almost certainly become scarce and therefore prohibitively expensive some time in the present century. Coal and nuclear will last several hundred years and so do not pose such an acute supply problem. The choice between them depends on other criteria such as pollution and climate change in the case of coal. In this chapter we consider coal, oil and natural gas; nuclear is considered in Chapter 4.

2.2. Coal

Coal is very plentiful. It made possible the industrial revolution and will remain a major source of energy for the foreseeable future. In the nineteenth century large cities like Birmingham and Manchester in the Britain and Pittsburgh in the USA grew up near the coalfields and millions of rural workers were drawn in to work in numerous factories. Raw materials were brought from faraway countries by coal-powered ships and the manufactured products exported all over the world. Huge railway networks were rapidly built and steam engines fuelled by coal transported people and freight with an ease and speed never before attained.

Coal varies in quality from anthracite that is almost pure carbon to the low-grade lignite coal. In many places the coal seams are at or near the surface, but as these become exhausted deeper and deeper mines became necessary. Originally the coal was cut by miners with pick and shovel, and still is in some countries, but in others sophisticated machinery does the work of many men. The technology is well-known and accepted, and the coal fire, merrily burning in the living room, became an essential part of the family home.

Although huge quantities of coal have been used for over a hundred years, there are still vast deposits in many countries, particularly the USA with 22% of world resources, Russia (17%), China (13%) and India (10%) (Avery 2007). It was estimated in 1996 that the coal reserves amounted to over a trillion tonnes or 760 TWy. At the present rate of world consumption, this is sufficient for well over two hundred years (see Figure 1.2). In the United Kingdom alone, it was estimated in 2005 that the reserves and resources in known mines amount to 500 million tonnes, the estimate for opencast mines is about 600 million tonnes, and for underground gas about 7 billion tonnes (MacKay 2008). This may be somewhat reduced because the rising cost of oil may make it economic to replace it by converting coal to liquid fuel. In Britain, however, coal production is declining and in 2006 coal imports were only 73% of consumption.

Nevertheless, coal has serious disadvantages. Coal mining is dangerous, dirty and unpleasant, and many miners contract debilitating diseases such as silicosis. More important for the whole population is the atmospheric pollution caused by coal burning. Until the prohibition of domestic fires, large cities suffered polluted air. In certain atmospheric conditions a deadly smog can form, stopping all transport except underground trains.

The pollution due to a coal power station depends on the quality of the coal, but in a typical case a coal power station emits each year about eleven million tonnes of carbon dioxide, a million tons of ash, half a million tonnes of gypsum, 30,000 tonnes of nitrous oxide, 16,000 tonnes of sulphur dioxide, a thousand tonnes of dust and smaller amounts of other chemicals such as calcium, potassium, titanium and arsenic. Of these the most harmful pollutant is sulphur dioxide. A study by the National Academy of Sciences shows that the sulphur dioxide released by a coal power station causes annually about 25 deaths, 60,000 cases of respiratory diseases and $12 million in property damage (Cohen 1977). To produce one gigawatt of electricity requires about 3.5 million tonnes of coal, and this contains about five tonnes of uranium. Most of the solids are trapped by filters, but a few thousand tonnes of ash escape into the atmosphere, carrying with it a corresponding fraction of the uranium. This accounts for the radioactivity emitted by coal power stations. The waste is poured into the atmosphere and damages our health, and safe disposal of the millions of tonnes of solid waste has to be arranged. The lungs of rural people are clean, those of city dwellers are grey and those of miners are black.

The emission of carbon dioxide from fossil fuel plants may be greatly reduced by pumping it into underground reservoirs such as pumped out oil wells, coal beds or deep porous rocks like sandstone that are capped by an impermeable layer of another rock. This process is known as sequestration or carbon capture (Liang-Shih Fan and Fan Xing Li 2007). It can reduce the carbon dioxide emissions into the atmosphere by up to 80% and is being studied by eight leading energy companies, and trials are under way in Japan, Norway and Canada. This is a promising development in principle, but it has still to be shown to be economically practicable. A report of the Royal Academy of Engineering gives an estimate of £30 per tonne of carbon dioxide. It would therefore be expensive to sequester the seven billion tonnes of carbon dioxide released every year, and this would increase the cost of electricity by at least one-third, possibly by 75%. Another estimate is that it may increase the cost of electricity to about 5.4 p/kWh. In addition, the generating efficiency would decline by about 10% and the capital costs would be very substantially increased. The cost is partially offset if the carbon dioxide is pumped into an oil well, as the increased pressure pushes more oil out of the well. The process, called carbon capture and storage (CCS), needs about 25 years of research and development at a cost of $20B. A proposal to build a CCS plant at Mattoon in Illinois was abandoned in December 2005 because it was estimated to cost $1.2B.

A serious problem of sequestration is the security of storage. If the carbon dioxide is stored under an impermeable layer of rock, there may still be cracks and fractures that allow some of the gas to escape. It is not possible to control or monitor any leakage that takes place. Even quite small leaks may be dangerous, and a large and sudden release can have serious consequences. Carbon dioxide is not in itself poisonous, but in large quantities it can be lethal if it displaces the air and so deprives living organisms of essential oxygen. In this way it can damage vegetation and asphyxiate animals and human beings.

The dangers of carbon dioxide have been shown by some natural releases. Thus 'in 1989 a slow release, at about 300 tons per day, was identified at Mammoth Mountain in eastern California from a geologically young dormant volcano. Levels of carbon dioxide at about 1% by volume killed about a hundred acres of forest. Although below the 10% level noxious to humans higher concentrations of up to 80% were found in enclosed spaces such as tents or cabins'.

'A more dramatic catastrophic event occurred on August 21, 1986 with the sudden release of about 1.6 million tonnes of carbon dioxide from a volcanic lake at Nyos in Cameroon. About 1700 people, mostly rural villagers as well as 3500 livestock, were asphyxiated. Most of the victims died in their sleep. The gas killed all living things within 15 miles of the lake. About four thousand people fled the area, and many of them developed respiratory problems.' It is not yet known how this happened. Some geologists think that it was due to a landslide, others think that it was due to a small volcanic eruption on the bed of the lake, while others again suggest that cool rain on one side of the lake triggered the event. 'Whatever the cause, the event resulted in a rapid mixing of the supersaturated deep water with the upper layers of the lake, where the reduced pressure allowed the stored carbon dioxide to effervese out of solution'. Pure carbon dioxide is denser than air, so it 'flowed off the mountainous flank on which Lake Nyos rests and down two adjoining valleys in a layer tens of metres deep, displacing the air and suffocating all the people and animals before it could dissipate' (Nuclear Issues 30, March 2008).

In the UK alone the coal power stations produce about 160 million tons of carbon dioxide each year. The consequences of a natural release of just 1.6 million tons in the Cameroon naturally give rise to some concern about the wisdom of sequestration in any places near human habitation.

Another method of removing some of the carbon dioxide is by using an amine scrubber. The amines combine with the carbon dioxide and the resulting liquid is then separated, the carbon dioxide going to storage and the amines are recycled. This process removes less than two-thirds of the carbon dioxide (Pearce 2008).

The total emission of carbon into the atmosphere from fossil fuel (coal, oil and gas) power stations is now about seven billion tons per year, rather more than one ton per person. The actual amount for a particular person depends on his or her lifestyle. Thus, for example, one round trip by a long haul flight emits about half a ton of carbon per person.

The carbon dioxide emitted is partially responsible for the global warming discussed in Section 7.3 and all the other chemical wastes pollute the atmosphere, the land and the sea. These disadvantages are so serious that our dependence on fossil fuels must be reduced.

After a detailed discussion Maslin (2004, p. 143) concludes that 'from the safety and environmental perspectives, the storage of carbon dioxide underground and/or in the ocean is really not feasible however helpful this would be in the short term'.

2.3. Oil

Oil has been known since ancient times, but it was only in the nineteenth century that it began to be used on a large scale. The first oil well was drilled in Pennsylvania in 1859, and thereafter production rose rapidly. Oil has a higher calorific value than coal and is more easily extracted and transported, and so during the twentieth century its use increased rapidly until it became one of the world's leading energy sources. It can be burned like coal for domestic heating and ship propulsion but is mainly used in large amounts in cars and aeroplanes that cannot be driven by coal. Thus in 2006 transport consumed about 60–70% of the oil. Huge oil tankers bring oil daily to refineries and distribution centres, whence it goes to petrol stations and airports. Our lives thus depend much more directly on a continuous flow of oil than on the availability of coal. If the oil supply is interrupted even for a few days there are immense repercussions as transport and industry come to a stop. In addition, oil is the basis of a large range of new petrochemical industries such as plastics, dyes, drugs, and paints. It can also be used for desalination of seawater in arid countries bordering the sea.

Oil wells generally last for about twenty or thirty years, and then the flow diminishes and it becomes increasingly expensive to pump it out. There comes a point when the energy spent in extracting the oil is greater that obtained by burning it, so that further extraction is uneconomic. Using sophisticated seismic and other geophysical techniques, oil companies therefore continually look for new oil fields to replace those that are becoming exhausted. Those found in recent decades in places such as Alaska and the North Sea are in much more inaccessible places than those in the Middle East, and the cost of extraction is correspondingly high.

The rate of oil production in the future can be estimated from the rate of discovery of new oilfields. The chilling fact is that the rate of discovery is falling, as more and more of the earth is explored. No new supergiant oilfields such as those in Alaska, that supply the bulk of the oil, have been found since 1975. It has been estimated that to maintain present oil production it is necessary to find new oilfields equivalent to that in the North Sea every two years. Geologists familiar with the poor rate of discovery during recent years believe this to be impossible even with the latest technology, so that oil production must soon reach its maximum and start to fall. In the USA, for example, oil production peaked in the late 1970s with the production of about 10 million barrels per day and by 2005 this had fallen to about five million barrels per day. Oil production in Venezuela peaked in 1998, in Indonesia in 1999 and in the UK also in 1999. The volume of newly-discovered reserves peaked in 2000 (Monk 2005). Worldwide, the peak is likely to be a broad one, and estimates of when this will happen vary. Thus IEA (International Energy Agency) predicts 2020–2030, BP (British Petroleum) 2015–2020 and ASPO (Association for the Study of Peak Oil and Gas) <2010 (Nuclear Issues 28, August). After the peak there will be a slow decline by about 3% per year, and as soon as the peak is passed the price will rise rapidly.

The European Community is heavily reliant on imports of oil and gas. A Report in 2001 found that 51% of oil imports come from OPEC countries and 42% of the natural gas from Russia. The oil from PEC countries was supplied by Saudi Arabia (13%), Libya (10%), Iran (9%), Iraq (7%) and Algeria (4%). Many of these sources are unreliable on the long term. The natural gas from Russia comes through a long pipeline that is politically vulnerable (Nuclear Issues 23, March 2001).

At present we are highly dependent on oil, particularly for transport. In addition, there will be increased demand from large developing countries such as China, India and Brazil.

The vital question is how long will there be enough oil, and the other fossil fuels gas and coal, to supply our needs. The current estimates are that at the present rate of use there is enough oil to last forty years, natural gas sixty years and coal 230 years. These figures are not so alarming as they appear, because they are obtained by dividing the known reserves by the annual consumption and this does not imply that after these times the reserves will be exhausted. Indeed, continuing studies reveal the surprising fact that these figures remain almost constant from decade to decade, as shown in Figures 1.1 and 1.2. The explanation is that as the existing reserves are used up the price rises and this stimulates searches for new oilfields and the development of new techniques for extracting more oil from existing ones. This produces more oil, so the price falls again. This in turn increases consumption, so that more oil is used and the price rises again. The overall result of this feedback mechanism is that the oil price remained fairly steady for some years in the range of $15 to $30 per barrel. Subsequent events have confirmed that this is over-optimistic and obviously it cannot go on forever.

In addition to these economic considerations, oil prices are subject to political decisions by the OPEC countries. This was the reason for the sharp rise in oil prices in 1973. Increased prices also occur when the demand exceeds the capacity of the refineries. At present average oil prices continue to increase: during 2004 and 2005 they rose from $35 to $55 per barrel, in 2008 they reached $120 and are predicted to rise still further.

These remarks refer to the world as a whole. The changes are more rapid in individual countries. Thus for example in Britain the oil will be exhausted in about five years and gas in about seven years. After that, without a new energy source, we will have to rely on gas imported from Libya and Russia.

Continuing dependence on oil is thus politically unwise. Most countries do not have their own oilfields, and so have to import the oil they need, putting themselves at the mercy of the oil-producing countries. Well over half the remaining proved oil reserves, amounting to between one and three trillion barrel or 130 to 320 GtC, are in the Middle East. In 1989 the percentages of oil imports from that region were 63% (Japan), 29% (Western Europe) and 11% (USA). This dependence is particularly critical because the Middle East oil reserves are expected to last about 100 to 150 years, whereas those in the rest of the world are likely to be exhausted in ten to fifteen years. Thus we will become increasingly dependent on Middle Eastern oil (Avery 2007, p. 111). By 2030 about 46% of world oil production is expected to come from the Middle East.

There are very large deposits of oil and tar sands in northern Alberta. It is however very costly to extract the oil, and it has been estimated that about two-thirds of the oil is consumed in the extraction process. There is also a large deposit of super heavy oil or tar in Venezuela. These sources of oil are heavily polluting, and if nevertheless they are developed will prolong the availability of oil, but not prevent its ultimate decline.

Thus while there is no reason to expect an imminent shortage of fossil fuels, there is a continuing need for flexible planning and the search for new sources. Despite its many advantages, it is imperative to reduce our demand for oil. It is extremely wasteful to burn it in power stations, especially because it is the only practicable source for aeroplanes, cars and the petrochemical industries. In addition, burning it in power stations produces carbon dioxide and so it contributes to global warming.

A proposal to alleviate the oil crisis has been made by the Association for the Study of Peak Oil. Their Oil Depletion Protocol 'requires the oil importing nations to reduce their imports by an agreed yearly percentage, the World Oil Depletion Rate, so as to put demand in balance with the declining world supply. At the same time the producing countries would agree to reduce their rate of production by a National Depletion rate — determined individually for each producing country as the total yet-to-produce oil divided by the yearly amount currently being extracted. The Department of Trade and Industry's figures for the UK continental shelf estimate the total remaining oil reserves at 1267 million tons with an annual production in 2005 of 85 million tons to give a depletion rate of 6.7%' (Nuclear Issues, April 2007). There will certainly be strong opposition to this proposal, but if it is accepted it will postpone the oil crisis, hopefully long enough for alternative energy sources to be developed.

Since a large fraction of the oil that is produced is used for transport, it is imperative to find new ways of driving our cars, buses, lorries and ships. Several possibilities are the subject of current research. One is to use liquid hydrogen as a fuel (see Section 3.10), another possibility is compressed air. Electricity is already used to drive trains and trolley buses, but cars require efficient ways of storing electricity. Already the Lotus company has developed the Tesla Roadster, a battery-driven car that does 135 mph, accelerates to 60 mph in four seconds and runs for 225 miles on one battery charge. However, it costs £50,000. The Indian company

Tata is working on a car driven by compressed air. The Italian company Pinifarina that works with Ferrari has built a car using hydrogen and fuel cell technology. These new developments are not yet commercially viable, but they indicate the paths that must be taken. Hopefully their costs will be reduced sufficiently by further research.

2.4. Natural Gas

Natural gas is often found not only associated with oil, but also independently. It can be used for heating and lighting, and before the advent of electricity gas was obtained from coal for this purpose. It is still widely used for domestic cooking and heating. Gas can be transported over large distances by pipeline, but in many places this is not practicable and so it is just burned at the oil wells and refineries. It can also be liquefied and transported in refrigerated tankers by ship, road and rail.

Gas can also be burned in power stations just like coal, and can also be used in many chemical industries. The large gas fields in the North Sea have made gas the cheapest form of energy at the present time, and hence stimulated what is called the 'dash for gas'. In Britain all the new power stations are gas-powered and they cost £400 per kWh to build. This is relatively cheap and convenient while it lasts, but it is predicted that the North Sea gas will be exhausted quite soon. After that, gas will have to be imported, with the same political dangers as oil.

The contribution of gas to world energy production has risen rapidly from less than 10 EJ in 1963 to about 80 EJ in 1993, accounting for about 21% of the world total. The proven reserves of gas are similar to those of oil, and the rate of consumption is only a little over one-half of that of oil. As the rate of consumption is rising rapidly it is unlikely to last longer than oil. A large gas field in Siberia is now supplying Western Europe with gas through a 5000-mile high pressure pipeline. In 1991, this supplied 20% of West European gas, including Finland (100% of total domestic supply), Austria (76%), Germany (34%), France (31%) and Italy (29%).

In the period 1992–1996 the total gas consumption in Britain increased by 45%, and this has led to price increases. The Energy Advisory Panel has estimated that the price of 17.3 pence per therm in 1990 is likely to rise to between 22.0 and 28.8 by 2005 and 25 and 37 by 2020.

The estimated natural gas reserves in 1996 amounted to 1412 trillion cubic metres, corresponding to an expected lifetime of just over sixty years. Of these reserves, 40% are in the former Soviet Union and 32% in the Middle East. For Britain, the estimated lifetime is about seven years. It is expected to peak a few years after the oil, and then rapidly decline.

In many respects, gas is now the cheapest, safest and most convenient energy source, but its lifetime is severely limited.

References

Lian-Shih Fan and Fan Xing Li (2007) Clean coal, *Physics World* (July), p. 37.

Maslin, Mark (2004) *Global Warming* (Oxford: Oxford University Press).

Monk, David (2005) *Physics and the Hunt for Black Oil*, *Physics World* (July), p. 37.

Pearce, Fred (2008) Cleaning up coal, *New Scientist* **29** (March), p. 36.

Chapter 3

The Renewable Energy Sources

3.1. Introduction

The term 'renewables' covers a variety of energy sources that are renewable in the sense that unlike coal and oil they do not use up the raw material. This is not exactly true because they all rely eventually on the sun, which provides the energy by burning fuel that will ultimately be exhausted, and in addition they use up material in order to construct the turbines, windmills, solar panels and other machines. There are several types of renewable energy, and among them hydro is a well-tried source. There are also the sources that rely on burning organic material, and finally wind, solar and several other possibilities that are listed in Table 3.1. This shows that only 3.6% of our electricity is obtained from the renewables, and if hydro is omitted this falls to 2.1%. These renewable sources will now be considered.

With a few exceptions, the renewables generate electricity, which accounts for only about one-fifth of our total energy consumption. The exceptions are geothermal and solar heat when they are used directly. Thus if we obtained all our energy from renewable sources we would still have to find a way to supply the remainder of our energy needs. To some extent this can be done without difficulty, but other applications require technologies that are still being developed. Thus heating that relies on fossil fuels can easily be replaced by electric heating, but transport is more difficult. Trains can be driven electrically, and cars suitable for short journeys can be driven by batteries that are recharged when necessary. It is practicable to use hydrogen instead, as described in Section 3.10, but it is at least initially likely to be more expensive, and the

Table 3.1.   UK Energy Generation 2001 in GWh (Times Higher, March 28, 2003)

Wind	967
Solar Photovoltaics	2
Hydro	4055
Landfill Gas	2507
Sewage Sludge	363
Solid Waste	948
Other Biofuels	770
Wastes	488
Total Renewables	10,099
Total for coal, oil and gas	279,213

hydrogen has to be produced by non-polluting sources. Ships can be driven by nuclear reactors. Aeroplanes pose difficult problems. Nuclear reactors and batteries are too heavy to be used, and the fuels currently in use are heavily polluting. Whether it is possible to use hydrogen remains to be seen.

It was announced in 2008 that '27 European countries had agreed to cut greenhouse gas emissions by 20% and commit the European Union to generate one-fifth of its energy from renewable sources within 13 years' (MacKay 2008). This is very optimistic, and it is not easy to see how this can be achieved.

3.2. Hydropower

Hydropower is often described as a renewable source, but it is so different in many respects from the other renewable sources that it is best treated by itself. It is a well-tried source that now accounts for about 3% of world energy production at prices comparable to those of the major energy sources. Renewables are expensive because the available energy, though very abundant, is spread over huge areas and has to be collected. In the case of hydropower the collecting is done for us by the river valleys. To ensure a steady flow of water and to increase the depth of fall, the rivers are dammed to form lakes. The water flows through the turbines to produce electricity. Hydroelectric power provides much of the energy in mountainous countries like Norway, Switzerland, Iceland and Austria.

Hydropower differs from the other renewables in being generally reliable and relatively rather dangerous due to the possibility of dam bursts. It cannot be significantly expanded because most of the suitable rivers in the developed countries have already been used. There are several possible sites in the developing countries, but these are usually far from centres of population and so are vulnerable to guerrilla attacks. Saboteurs have often blown up the pylons carrying electricity from the Calboro Bassa dam on the Zambesi river. Hydropower occupies large areas of land and ultimately the lakes silt up and are no longer useful. Severe problems have been experienced in several countries heavily dependent on hydropower (Norway, New Zealand, Brazil and the USA) during periods of prolonged drought.

Hydropower inevitably inundates large areas of land, often including villages and fertile agricultural land. It can also disrupt the ecosystems of the surrounding areas. The dam prevents the upstream migration of fish such as salmon, though this may be overcome by building fish ladders. The areas below the dam are deprived of silt that is valuable for the soil; this was an important disadvantage of the Aswan dam on the Nile.

In a few places there are areas of land below sea level that could be used to generate electricity hydroelectrically from sea water flowing into them. It has been estimated that 4000 MW could be generated in this way using the Quattara depression in North Africa.

3.3. Wind

Windmills have been used to grind corn since ancient times, and they are still widely used to pump water from wells on farms. In such cases it does not matter that the wind is not always blowing, because the water can be stored in reservoirs above ground until it is needed. Windmills can easily be connected to a generator to produce electricity. Modern windmills are not like those in pictures of Dutch rural scenes; they are large propeller blades driving turbines mounted on high towers. About a thousand of these, spaced half a mile apart, are needed to equal the output of one coal power station. They are unreliable because the wind velocity is very variable, and the electricity they produce is relatively expensive. The energy output of a wind turbine is even more variable, since it is proportional to the cube of the wind velocity; this means that at low velocities very little power is produced, while at high velocities the blades have to be feathered to prevent damage. The result of this is that the average efficiency, or load factor, of a wind turbine is only about 25% of its maximum value. The output can change rapidly; for example the power from a 6000 MW array of windmills in Germany once fell by 3600 MW in six hours. Such fluctuations make it necessary to have a large back-up capacity able to take over at short notice. This must be kept on standby, which uses fuel and causes emission of greenhouse gases. Large arrays of wind turbines, called wind farms, have been built in several countries, but these require substantial subsidies. As the cost of this becomes evident, many wind farms are being abandoned.

Wind has always been used for ship propulsion and continues to be important for fishing boats and yachts. Large ships require coal or oil, but even for them some energy may be saved by adding sails for use when the wind is favourable.

Apart from hydro, wind is the most promising of the renewable sources, but windmills are certainly more costly and unreliable than the major power sources. It is not easy to estimate the cost of wind power because so few wind turbines have operated long enough for reliable statistics to be obtained. The technology is relatively new and many new designs are being studied.

In spite of intensive efforts, the contribution of wind power is still just a few per cent, as shown in Table 3.2 for the year 2001. The world total was then about 13.4 GW. By the following year, the wind power capacity had increased in several countries: in Germany to 6 GW, in the USA to 4.2 GW, in Spain to 3.3 GW and in Denmark to 2.5 GW (SC January 2002). It is notable that there are no figures available for Britain, although some have now become available. Denmark is ideal for wind power but short of other energy sources, and wind contributed about 10% of the total energy, rising to 18% in 2005 and 23% in 2007. Denmark has an arrangement with Sweden to exchange energy according to their needs. In 2005 Denmark exported 55% of its wind energy at a low price, and at times of need had to buy energy back at a higher price. This reduces the contribution of wind in Denmark to about 8.5%. The Danish Government has now decided that an expansion of its renewables programme would be the most expensive option, and that it would produce no overall reduction in carbon dioxide emissions (Nuclear Issues 25, March 2003).

Table 3.2. Wind Power Performance in European countries (IAEA Wind Energy Report for 1999. Nuclear Issues 22, July 2002)

Country	Capacity MWe	Output GWh	Load Factor %	Share of Electricity %
Denmark	1752	3055	19.9	10
Finland	36	49.3	14.8	0.06
Germany	4445	7400	19.0	1.7
Greece	107	160	17.1	0.25
Italy	282	400	16.2	0.15
Netherlands	416	688	18.9	1.0
Norway	13	25.4	22.3	0.03
Spain	1539	3750	27.8	1.6
Sweden	215	365	19.3	0.25
UK	344	NA	NA	NA

Wind farms occupy large areas of land, so many of them are now being built off-shore. These have the advantage that they do not incur land costs, but have the disadvantages of being more difficult to construct and maintain, and underwater cables have to be used to transport the electricity to the land. They are also a potential danger to shipping. The first off-shore wind project in 1990 was a 220 kW turbine in Sweden situated 0.25 km from the shore, and this was followed by a 4950 kW wind farm off the Dutch coast.

Two 90 m high 2 MWe wind turbines have been built in the sea off the Northumberland coast. They cost £2M each and about 1300 of then would be needed to supply 1.8% of Britain electricity needs. There are plans in Germany to build 2000, 5 MWe wind turbines in the Baltic Sea. Ten times that number would be required to equal the electricity produced by their nuclear power stations, and they would occupy all German coastal waters out to a distance of 6 km (Nuclear Issues 23, February 2001). The German company E. On plans to build a wind farm of 300 MW with 83 turbines off the Yorkshire for a cost of £700M. The Ministry of Defence is objecting because the blades can create clutter on radar screens, thus interfering with the radar defences.

Globally, wind power grew by 30% in 2001 and by 2002 the total capacity had grown from 17.8 GW to 23.3 GW. For particular countries, the capacity of Germany became 6 GW, Spain 3.3 GW, Denmark 2.5 GW and the USA 4.2 GW (Speakers' Corner No.141, January 2002). By 2006 the United Kingdom output reached 1935 GWh, less than 0.5% of UK electricity (Nuclear Issues 28, April 2006).

A large wind turbine near Aalborg has a power output of 630 kW and cost £500,000; smaller ones generating 30 kW cost £10,000. Another estimate is that 1 GW costs about one billion pounds. Recently the British Government decided to build several offshore wind farms by 2020 producing 33 GW at an estimated cost of £40B; subsequently the estimated cost rose to £80B. These costs are so high that a wind programme is only possible with a Government subsidy. Since backup power stations must be built to take over when there is no wind, and since these are much more economical than wind power, one may question why wind turbines are built at all.

In many countries the subsidies for wind power have been cut or cancelled, stopping further development. Thus in 2002 the Danish Government withdrew subsidies for three new offshore wind farms. The newspaper Jallandsposten commented that the wind turbine industry is dependent on wind power subsidies: 'wind power is unreliable, very expensive, badly scars the landscape and is only useful as a supplementary source of power, nothing more'. Quite recently, Royal Dutch Shell have withdrawn from the planned London Array wind farm because 'the costs of wind farms are spiralling out of control. The price of steel and turbines is soaring. Ships needed to install turbines are booked out for years to come. Siemens, which makes the turbines, has no spare capacity (Daily Telegraph, 2 May 2008, p. B4).

In the year 1999, 318 GWh was generated in the UK by windmills of rated capacity 745 GWh, a load factor of 26.7% (Nuclear Issues 22, July 2000). German wind generators in 2001 had load factors of less than 20% and produced 7.5 billion kWh per year, less than 2% of consumption. To replace nuclear generation would require 100,000, 1 MWe wind generators (Nuclear Issues 23, 2001). Globally, wind produces about 0.06 TW per year compared with a total output of about 11.2 TW, just about 0.5%.

Wind turbines have to be sited on high ground in windy locations, and there is increasing opposition on environmental grounds. For example, it is proposed to build a wind farm of 27 turbines each 400 ft high in the Lake District, an area of outstanding natural beauty. This was opposed by the Campaign to Protect Rural England because it would adversely affect the landscape, but supported by Friends of the Earth because it would reduce carbon dioxide emissions.

Many of these turbines have been built and opened with a fanfare of publicity, only to be quietly forgotten a few months later when the propeller snaps or blows away. For example, an experimental 3 MW wind turbine at Growian in Germany was abandoned after operating for less than two years. During this time it produced 80,000 kWh, enough to meet the needs of 23 households, at a cost of about a million pounds. In 1987, a 3 MWe wind turbine with blades 60 m in diameter was erected on the island of Orkney. It was expected to generate 9 GWh/y but it failed to operate satisfactorily and was dismantled in 2000.

The British Government now subsidises wind power by about £20–25/MWh, which may be compared with the price of £15–18/MWh paid to France for nuclear electricity. The total cost of these subsidies up to 2020 will amount to about £B30 (Nuclear Issues, June 2004). Even with these subsidies, wind provides less than 0.25% of Britain's energy needs. The Government is committed to achieving 10% by 2010, which require about 8400, 2 MWe turbines, assuming 25% efficiency. These would have to be spread over an area of about 1300 square kilometres, more than the Isle of Man and the Isle of Wight combined. Even to achieve the contribution of 6 or 6.5 GWe believed to be possible by the British Wind Energy Association would require 3000 to 3200 of the largest turbines. Only 61 turbines with a total capacity of 103 MWe were built in 2003 (Nuclear Issues 26/7, 2004).

To produce ten billion kWh each year would require 2400 giant offshore wind generators of 2 MWe occupying 380 square km. To supply all the energy needs of the UK would require 84,000 of these windmills and the output would be unreliable. The same amount of energy could be provided reliably by 35 nuclear power stations of 1300 MWe, each occupying 0.2 square km.

In 2005, 2908 GWh was generated by windmills in the UK, compared with a total energy generation of 408,846 GWh, just 0.7% (Nuclear Issues 22(8), September 2006). The goal is to raise this to 1.8% by 2010.

A serious difficulty with wind power would arise if it ever provides more than small amounts of electricity. Due to the unpredictable variability of the wind strength, the amount of electricity also fluctuates, and this makes it difficult to integrate it into the electricity grid. A possible way to overcome this is to store the energy by using the windmills to pump water from a lower lake to a higher one. The fluctuations in production are then not important, and the energy can be used when it is needed. This solution is only practicable where there are suitable lakes, and these are rather rare. It is also possible to use tidal basins and hydroelectric power stations to store wind energy. The growing shortage of oil is likely to make it economic to use electric cars, and these could store wind energy produced at irregular times and even supply it in times of need (MacKay 2008, p. 157). All these innovations could do much to overcome the disadvantage of the intermittency of wind power.

The maximum energy from wind power for each person in the UK can be estimated from the power of each windmill, the area it occupies and the population density. Making extreme assumptions, such as that windmills occupy 10% of the country. MacKay (2008) finds it to be 20 kWh per day. This cannot be increased by increasing the power of each windmill or by building them closer together because the more powerful ones require more space because if they are too close together they shield each other. For a particular power there is a minimum spacing, and this increases with the power. The more powerful they are, the further apart they must be.

Wind power is particularly useful on remote windswept islands and mountainous regions where the electricity demand is insufficient to justify a large power station. Electricity in such places is usually provided by a diesel generator, and this can be kept as a backup for use when the wind fails. Such a system is operating on the Norwegian island of Froya, and consists of a 55 kW wind turbine combined with a 50 kW diesel generator.

The Global Wind Energy Council predicts that by 2030, 29% of world electricity will be produced by wind. The International Energy Agency predicts a contribution of 3.5%.

3.4. Solar

The sun pours energy on the earth at the rate of about 100 PW/yr, equivalent to $5 \times 10 (20)$ J/y, enough to satisfy our energy needs about seven thousand times. If we could find a practicable way of using it we would have solved the energy crisis. The heat of the sun stirs up the atmosphere and causes winds and waves, and its energy can be used to generate electricity or used directly. On average, the sun gives 1400 watts per square metre to the earth and this is reduced by clouds to about 200 watts per square metre at the earth's surface, with higher values of up to 1000 watts per square metre in desert areas.

The energy in the sun's rays can be captured in several ways. The first way is to use photovoltaic cells to convert it directly into electricity. In these cells the solar photons are absorbed by a semiconductor and release electrons to form an electric current. The photocells generate a rather small voltage, so to increase this to a useful level it is necessary to have thousands of them in series. Photocells have losses that reduce the energy by a factor of about four so that the average irradiation is about enough to light one ordinary light bulb from the sunlight collected by one square metre. The area of a collector required to satisfy the energy needs of an average household had been estimated by Professor Hoyle to be about 1200 square metres. Thus a collector about the size of the Jodrell Bank radio-telescope would be required for each group of four families. To produce the same amount of electricity by a modern power station would require collectors covering about 50 to 100 sq. km.

The cost of photocells is so high that it is not practicable to use this method for large-scale energy generation; it would only become economic if the cost could be reduced by a factor of about ten, quite apart from the value of the land occupied. If remote desert areas are used, construction, maintenance and transmission costs are increased. There are many optimistic projections into the future. Thus Crabtree and Lewis (2007) estimate that the cost will be reduced to 2 c/kWh in 20 to 25 years, and that by 2050 photocells will supply 25% of the world's energy.

Although solar electricity is still too costly for general use, there are some applications where solar energy is extremely useful. These occur when a relatively small amount of electricity is needed in places where it is impossible or uneconomic to provide it in any other way. Examples are road signs in remote desert areas and artificial satellites.

The second way to harness solar energy is to focus the sun's rays on a central boiler and to use the steam produced to drive a turbine to generate electricity. A solar power plant in California uses this method to heat molten salt to 566°C and generates 10 MWe (Avery 2007, p. 136). Alternatively, a parabolic mirror can focus the sun's rays on a central pipe carrying the liquid to be heated. The liquid can also be oil. To maintain efficiency all the mirrors have to be kept clean, and must be controlled by servomotors to keep the sun's rays focussed on the boiler as the earth rotates.

Because of the high cost, the contribution of solar energy remains very small, less than 0.1% in the USA in 2004.

A simpler but less flexible way to use solar energy is to use it for domestic water heating by putting a solar panel on the roof of a house and letting the sun's rays heat directly the water running through a grid of pipes. Aided by government subsidies, many houses are now equipped with these solar panels. On sunny days this is enough to keep the domestic water hot, but a normal boiler is needed as well to boost the water temperature when necessary.

Houses can now be designed to catch and retain as much as possible of the sun's heat. South-facing windows and wall insulation can greatly reduce heating requirements. It is less easy to modify existing houses, but much can be done by insulation and double glazing.

Solar ovens, with a large mirror to focus the sun's rays, are useful for cooking meals in sunny countries. While this is perfectly practicable, it means that meals can best be cooked at midday, and not in the evening, which is the traditional time in many countries such as India.

3.5. Tidal

The tides can be used to generate energy by damming a river estuary and generating energy by letting the water flow through turbines when it comes in and also when it flows out again. A small 240 MW tidal power station with an average energy output of 65 MW has been operating for many years at La Ranche in France, but it is not economic. A detailed study has been made of the feasibility of using the Severn estuary between England and Wales; which is much larger and could produce up to 7 GW. It is estimated that the energy it would produce would cost about twice that of a conventional power station. An investment of about fifteen billion pounds spread over about ten years would be needed before any power is produced, and the environmental effects are likely to be severe. Tidal power is reliable but periodic, and the power is not always produced at the time of day when it is needed, since tides follow the lunar cycle.

The intermittent nature of tidal power can be overcome by having two basins behind the tidal barrier. When the tide is high it flows into the first basin, generating energy. When the level in the basin is the same as that of the sea, the water is allowed to flow into the second basin, generating more energy. When the tide goes out, the water in the second basin flows out until it is empty, generating more energy. The energy production is kept at a steady level during the turns of the tide by letting water flow from the full to the empty basin. The efficiency of the system can be increased by pumping more water into the first basin at high tide, and pumping it out of the second basin at low tide.

The choice of site is important for tidal power because the height of the tides varies greatly from one location to another. An extreme example is the Bay of Fundy between New Brunswick and Nova Scotia where the tidal waves are funnelled into the bay, resulting in a tidal range of 17 m.

To reduce the cost, a tidal power station must be built on a large scale, and suitable estuaries are rare. They take so long to build that a massive initial investment is required at a low interest rate. The operating costs are likely to be high due to corrosion of the turbines by sea water and the growth of barnacles and seaweed. It is thus unlikely that the such tidal power stations will ever become a useful source of energy.

Another possibility that is being actively studied is the use of underwater turbines in the English Channel or near the Orkneys or wherever the current exceeds about 5 knots. These would not have many of the disadvantages of tidal power stations and could generate 500 kW and be totally reliable. A very large installation is being built in the Cook Strait near New Zealand. It consists of 7000 turbines anchored 40 m below the surface to avoid storms. This could provide all New Zealand's electricity (Avery 2007). The British Government has recently approved a prototype tidal project in the Humber estuary. It is hoped that it will generate 0.1 MW for an expenditure of £878,000.

3.6. Wave

The energy in ocean waves comes from the winds, and this in turn comes from the sun. The energy given to the oceans in this way has been estimated to be 2.7 TW per year. The power transmitted by a wave varies as its period and the square of its height. In mid-Atlantic this reaches about 90 kW per metre of wavefront, for waves with a height of about 4 m and a period of about 10 s, but falls to less than 20 kW per metre on the coast due to friction with the seabed and by its breaking motion (Heath 2005). This implies that devices out at sea are able to tap more energy, but they are more vulnerable than devices on the coast that can tap less power. The energy in waves is quoted in kW per metre because any device installed at a certain distance from the shore would absorb the wave energy and prevent any further devices nearer the shore from obtaining any more energy.

Many devices have been proposed to convert wave energy into useable form, but so far none of them has operated commercially on a large scale. Inevitably the devices have to be very large and therefore costly, and they are subject to corrosion by sea water and buffeting by the waves. The £3.5M wave power generator Osprey 1 weighing 8000 tons sank in shallow water off the Scottish coast in a freak summer storm. It was designed to produce 2 MW of electricity, but never operated. More recently, a much publicised device, costing over a million pounds, is designed to produce 75 kW, enough to power 25 three-bar electric fires. An oscillating water column wave power device designed to produce 100 to 500 kW was built in Norway. It was expected to produce electricity at 3 to 4 pence per kW, based on a lifetime of twenty-five years, but after operating for three years it was destroyed by a storm (Ross 1995; Heath 2005).

Another device, known as LIMPET, 'consists of three inclined concrete shafts that dip beneath the surface of the ocean. A submerged opening at the foot of each shaft lets in water, allowing waves to alternately push water into and pull water out of the shaft. This motion in turn draws air into and out of the top of the shaft, powering a turbine'. The energy conversion efficiency was estimated to be 48% to give an average power input of 202 kW. In practice the mechanical and electrical energy losses reduced the mean average output to 21 kW (MacKay 2008). Since 2000 this device has been supplying power to the local electricity grid.

A wave power generator device, called Pelamis, is being tested off the Orkney Islands. It 'consists of four floating cylinders, each about 30 m long and 3.5 m in diameter, connected by hinged joints. The waves cause the cylinders to move relative to each other and this motion is used to pump high-pressure oil through hydraulic motors, which drive electric motors'. It is designed to withstand the powerful waves that are expected in the open sea. Another similar device is Salter's duck.

In Portugal, a Dutch company has installed a device that contains a volume of trapped air below sea level; as the waves pass over it the air is compressed and generates energy. There are many other wave power devices being made worldwide.

All these wave energy devices are presently uneconomic, but their designers are confident that the costs will eventually come down to acceptable levels. There are very few statistics available. Some Governments provide subsidies for wave power; for example the Portuguese government pays 16 p/kWh for wave-generated electricity. As this is about six times the current price it is clear that wave power has a long way to go before it becomes a practical proposition.

3.7. Geothermal

The interior of the earth is still cooling down from the time it was red hot, and additional heat comes from the decay of radioactive rocks. This heat is continually flowing outwards and can provide a source of energy. The amount of heat released varies from place to place, and in some regions it escapes easily through volcanoes and hot springs. The water from these springs is sometimes hot enough to be used as a source of energy, although more often it is used in medicinal baths and as a tourist attraction. Often the hot water is accompanied by toxic gases such as hydrogen sulphide. The use of geothermal energy in this form can be economically viable if large amounts of very hot water are readily available, or if it can be used for domestic heating.

Such regions are relatively rare, so if geothermal energy is to be used as a worldwide energy source the heat must be extracted from the earth's crust. On the average, the outward heat flow is about 50 mW per square metre, of which about 40 comes from radioactivity and the rest from the cooling of the core. To be useful, we need temperatures of at least 150°C to 200°C. The temperature of the earth's crust increases with depth by about 25°C per km. So the deeper we go the better, except that it becomes increasingly expensive to drill holes. The heat can be extracted in several ways, depending on the conditions. If the rock is dry and porous, two holes can be drilled some distance apart, and cold water pumped down one hole, driving hot water up the other. If the rock is wet, the hot water can be pumped up directly. After a time, however, the rock around the holes cools down and is no longer useful as a heat source. Particularly in the case of dry rock, it takes a very long time for it to heat up again by conduction from the surrounding rocks. The amount of heat that can be obtained from a particular hole is thus severely limited. It is very costly to bore such holes, and to ensure that they are the optimum distance apart. If they are too close, the amount of heat obtained is small, while if they are too far apart, the water pumped down one hole cannot get through to the other. Many experiments have been made to assess the economic practicability of this type of geothermal power, and the results are disappointing. Very large amounts of heat are there, but it is too expensive to extract it. Detailed calculations of the available geothermal heat are given by MacKay (2008).

A study of the feasibility of geothermal power in Cornwall was abandoned after an expenditure of £33M. The energy obtained was estimated to cost about four times as that from wind or solar. An application to build a geothermal power plant at Heber in California was refused because the Public Power Commission found that 'the power delivered would cost between 17.9 and 24.3 cents per kWh compared with 11 cents per kWh for a coal power stations and 16.6 for an oil one. In 1976 the total geothermal electric capacity amounted to 1325 MW, principally from hot springs. By 1990 this had increased to 6 GW, and it subsequently increased by 3% per year.

The economic use of geothermal energy is thus limited to the rather few places where hot water or steam escapes from the earth. The total amount of energy generated in this way is a few thousand MW, and so it is evident that geothermal power can make only a small contribution to world energy needs.

3.8. Ocean Thermal

There is an appreciable difference in temperature between the surface and the depths of the ocean and this can be used as a source of energy. In tropical regions the surface is at about 25°C and at about 1000 ft the temperature is 5°C, and this temperature difference is sufficient to drive a turbine. The feasibility of this process was first demonstrated in the 1930s by Georges Claude. More recently the idea has been developed by Abraham Levi and a 50 MW plant has been constructed off Hawaii.

The difficulties of these devices are the corrosiveness of the sea water, the problem of anchoring the device and transmitting the power generated to where it is needed.

3.9. Biomass

As already mentioned, wood is still extensively used as a major source of fuel in poorer countries. If more wood is used than is replaced by additional growth this can lead to desertification. It is therefore desirable to replace wood by more efficient and less damaging fuels, and certainly the use of wood cannot hope to solve the energy crisis in the developed countries.

Biomass in general is an organic material that can be burned to produce heat or allowed to decay and emit natural gas. In the form of wood and animal dung this has been done for centuries. In modern terms it is taken to mean use of the by-products of agricultural processes and the cultivation of plants or trees for the specific purpose of generating heat, and hence electricity. It is sometimes said that this does not add to global warming since the carbon dioxide produced when it is burned equals the amount absorbed by the photosynthesis that originally produced it. However, the House of Lords Select Committee report on 'EU strategy on Biofuels': from 'Field to Fuel' pointed out that 'carbon savings are affected by agricultural practice, production and processing methods and transportation of the feedstock'. They concluded that 'although biofuel use produces less carbon dioxide emissions than use of fossil fuels this may be partly, if not wholly, negated by environmental costs in their country of origin and by transportation to the point of use.'

An example of the former is the use of the straw remaining after corn is harvested. If this were all burnt it could provide as much heat as 2% of Britain's oil imports. As straw is cumbersome to transport, it is best burnt on the farms, and many farmers have installed straw boilers. The heat generated can be used to heat farmhouses and animal buildings and to dry grain and hay. However, the straw-burning boilers are expensive and require automatic stoking machinery. Most farmers prefer to burn it in the fields, where it fertilises the soil and kills weeds and insects. In Brazil the bagasse remaining after the sugar cane has been crushed is often burnt to heat the water used to purify the sugar or in gas turbine systems to produce electricity. Sweden and Finland obtain about 30% of their electricity from biomass, mostly from their forestry and paper-making industries.

Experiments have been made to determine the practicability of planting trees that are subsequently burnt. Willow is very suitable for this purpose, as it grows rapidly. The process has been found to be uneconomic, and it also uses up agricultural land.

In many countries, especially developing ones, there is a serious and growing shortage of food. Dominique Strauss-Kahn, the head of the International Monetary Fund, recently said that 'hundreds of thousands of people will be starving', and that 'children will be suffering from malnutrition'. 'He predicted that increasing food prices would push up the cost of imports for poor countries'. Since January 2007, the price of wheat has risen two and a half times and that of rice has almost trebled. These are staple foods in many countries, and the rising prices bear heavily on the poor. The sharp rise in food prices is due to increased demand, poor weather and an increase in the area of land used to grow crops for more biofuels. Of these, biofuels derived from soya beans, sugar cane and corn have been identified as the major cause. The food shortage and rising prices have led to riots in Egypt, Haiti, Bangladesh, India and several other countries. Professor Beddington, the British Government's Chief Scientific Adviser, has said that 'it is very hard to imagine how we can see the world growing enough crops to produce renewable energy at the same time meet the enormous demand for fuel' (Daily Telegraph, 14 April 2008). The food shortage could be considerably reduced by increased use of genetically modified crops, but these still encounter severe opposition.

The current massive switch from food production to growing crops for fuel forces us to answer the question: Do we really want people to starve so that we can continue to drive our cars? Thus according to the World Bank the price of food has risen by 80% in the three years to 2008, due to shortages of supply and rising demand, pushing 100 million people below the poverty line. This has partly been blamed on the USA and European Union for encouraging the conversion of corn to ethanol as an alternative to petrol. The European Union plans to use this for 50% of its transport fuel by 2010, and large increases are also planned in the US. However, it failed to reach its 2% market share target in 2005, but still aims for 5.75% market share by 2010 and 14% by 2020. The difficulty is that 'biofuels in Europe are expensive, between 30 and 45 p/litre whereas biofuel made from sugar cane costs 6–11 p/litre in countries such as Brazil that has higher crop yields and lower costs for land and labour' (Nuclear Issues, May 2007). A large-scale expansion of biofuel production also affects 'soil fertility, water availability, pesticide and fertiliser use'. The committee of the House of Lords already mentioned concluded that: 'Tropical rain forests act as a carbon sink: burning, logging and then ploughing it leads to very significant carbon emissions, so any potential benefit from growing cheaper renewable feed stocks on such cleared rain forest would never repay the carbon debt that you had built up by clearing it in the first place'.

The growth of biofuels requires large areas of land. Thus assuming that it gives 1.45 tonnes of fuel per hectare 29.5 million hectares would be needed to supply transport fuel for the UK, whereas the total available agricultural land is 5.7 million hectares. As James Lovelock has remarked, 'biofuels are potentially dangerous and if exploited on a large scale would lead to disaster' (Nuclear Issues, January 2007).

Biomass can also be used as an additional fuel in coal power stations. It is planned to adapt Britain's largest coal-fired power station, Drax, to burn every year about 1.5 million tons of biomass, such as olive stones and sunflower seed husks, together with the coal. It is estimated that this will reduce carbon dioxide emissions by 15%.

3.10. Hydrogen

Hydrogen can combine with oxygen to form water, releasing heat in the process. It can be produced by electrolising water, but since this requires more energy than is released by its combustion, hydrogen cannot be regarded as an energy source. It can however serve as a useful way of carrying energy from the generator to the point of use. Thus hydrogen could be produced by generating electricity by nuclear or renewable sources and then used to drive cars and trains, and possibly planes. Hydrogen is easily stored either as a gas or in solid metal hydroids, which can easily absorb up to a thousand times their volume. The whole process emits virtually no greenhouse gases. At present the technology has not been fully developed but if the difficulties can be overcome it could be a useful energy transporter in the future (Eikerling *et al.* 2007).

3.11. Costs

The acceptability of an energy source depends rather critically on the cost, and this is not easy to determine. It depends on many variables that vary from one country to another and on differences between the sources themselves. Among them is the cost of the raw materials; some, like wind and solar, are free, while in other cases the fuel has to be mined, processed and transported to the power station. Distribution costs depend on the distance between power station and consumer. Some power stations, such as gas, can be built relatively easily and cheaply, while others are more expensive and take longer to build. Some go on producing power for many decades while others have short lives. Furthermore, it is impossible to estimate in advance the life of a power station, and as it may last for decades the eventual cost depends critically on the rate of inflation and on the interest charged on the initial capital. Account also has to be taken of the cost of decommissioning the power station after it has reached the end of its useful life and returning the site to its original state. In addition there are the costs of the deaths and injuries associated with any large industrial activity. Even more difficult to quantify are the costs of the pollution of the atmosphere, the effects of acid rain, and the climate changes attributable to the greenhouse effect. Even without these latter costs, it is no wonder that there are considerable differences among the published figures of costs of the different energy sources.

It is particularly difficult to estimate the costs of the renewable sources. Many of the published figures refer to future projections made by the designers, not to actual experience, and in many cases these estimates have proved to be over-optimistic. It is essential to have cost estimates based on plants that have been running for several years, and these are difficult to obtain.

In the context of the energy crisis it is only useful to compare the costs of the major sources and of wind, the best of the renewables. The other renewables are so costly and unlikely to make any useful contribution to our large-scale energy needs that they need not be considered further.

A commission was appointed by the Belgian Government to examine the costs of energy generation in various ways. They took into account the costs of fuel, investment, operations and maintenance, atmospheric air pollution, noise, greenhouse gases, construction, grid connection and decommissioning and produced the estimates shown in Table 3.3. For comparison, some figures from the Performance and Innovation Unit, the Royal Academy of Engineering, British Nuclear Fuels a French and a Swedish source are also given. To facilitate comparison the Belgian figures have been normalised to equalise the costs of coal production. The RAE figures for gas were based on the current price of 23 p/therm; if the estimated forward price of 40 p/therm is used, the gas cost would rise to 3.2 p/kWh (Evans 2004). The fifth and sixth columns show figures including external costs. These have been estimated as 4–7 for coal, 1–2 for gas and 0.25 for nuclear. Taking account the different

Table 3.3.   Costs of Electricity Generation in p/kWh

Energy Source	Belgian	Costs*	PIU+	RAE"	BNFL**	Fr^	V^^
Coal	2.34	(3.5)	3.5	2.5–3.2	7.2	4.88	4.9–5.6
Gas	1.74	(2.6)	2.0	2.2	3.8	4.24	5.6–6.5
Wind (off-shore)	2.39	(3.6)	3.0	5.5 (7.2)	6	—	—
Wind (on-shore)	3.26	(4.9)	2.5	3.0 (5.4)	—	—	—
Nuclear	1.25	(1.8)	4.0	2.3	3.5	3.3	3.7–4.0

* Figures in brackets normalised to PIU coal (Nuclear Issues 23, February 2001).
+ Performance and Innovation Unit (Speakers' Corner, October 2001)
" Royal Academy of Engineering (backup costs in brackets)
** British Nuclear Fuels, including external costs (Nuclear Issues 23, September 2001)
^ French estimates in ecents/kWh, including external costs
^^ Vattenfall in cents/kWh (Nuclear Issues 29, December 2007).

circumstances there is reasonable agreement among these figures, except that the figure of 4.0 for nuclear for PIU seems definitely too high. There is also notable disagreement concerning whether off-shore or on-shore wind is the more costly.

In addition, Vattenfall estimates the cost of hydro to be 4.4–6.6 and biomass as 6–6.6, compared with a BNFL estimate of 4.5. The cost of photovoltaics has been estimated to be 25–60 c/kWh. The output of the wave power station on Islay is sold at 5.96 p/kWh (Speakers' Corner, December 2001).

Several estimates of costs have been made in the USA. McGraw Hill's Utility Data Institute found in 1999, 2.07 c/kWh for coal, 3.18 for oil, 3.52 for gas and 1.83 for nuclear (Nuclear Issues 23, February 2001). The costs of renewable sources were estimated by the National Energy Policy Group in the USA which found wind 4–6 cents/kWh, geothermal 5–8, biomass 6–20, solar 20 and hydro 2–6 (Nuclear Issues 25, February 2003). In the period from 1988 to 2000, the US production cost fell from 2 to 1 p/kWh (BNFL Report 2000).

At the end of its life all energy sources must be dismantled. The costs of dismantling are greater for nuclear because of the high level of radiation inside the reactor. In the USA, the Government simply charges $1 per MWh to be responsible for the ultimate disposal of spent fuel. Other countries have more complicated arrangements whereby a fraction of the estimated cost of decommissioning is set aside each year. In Sweden the decommissioning charge has fallen from 1.7 ore/kWh in 1982 to 0.6 ore/kWh in 2001 (Nuclear Issues 23, October 2001). The total cost of nuclear power is estimated to be 15–18 ore/kWh, equivalent to 1–1.3 p/kWh (Nuclear Issues 28, May 2006). The cost of decommissioning a Magnox nuclear power station can be met by an extra charge of 0.12. p/kWh on the electricity generated (Nuclear Issues 28, January 2006).

An extremely important aspect of the costs of energy from the various energy sources is their expected variation in the future. Oil prices fluctuate as a result of political decisions, but on the whole rise quite steeply. This rise will certainly accelerate when the oil production peaks and the demand exceeds the supply. Gas prices are rising so rapidly that any figures quoted at a given time are likely to be rendered obsolete before the estimates are printed. Just from 2003 to 2004 they rose by 42% to 25.2 p/therm. Coal prices are likely to remain steady because of the huge reserves. Nuclear prices are not significantly affected by a rise in the price of uranium because the cost of the fuel is a small fraction of

Table 3.4. Environmental Costs of Electricity from Various Sources in e/kWe (Nuclear Issues 23, August 2001)

Country	Coal	Gas	Oil
Belgium	4–15	1–2	0.5
France	7–10	2–4	0.13
Germany	3–6	1–2	0.2
UK	4–7	1–2	0.25

the running costs. They are likely to fall in the longer term as new reactor designs are brought into operation. Already the cost of electricity produced by British Energy fell from 2.37 p/kWh in 1995 to 1.67 in 2001 (Nuclear Issues 34, September 2002).

All these direct costs may be estimated quite well, but the indirect costs are likely to be far larger and impossible to quantify with any accuracy. The results of an attempt to do this are given in Table 3.4. These are the costs of the acid rain and climate change due to the burning of fossil fuels. Oil would be far more expensive if the costs of the wars fought for it were taken into account.

More extensive estimates including health effects, noise nuisance, acid rain, global warming and climate change have been published by the Nuclear Energy Agency News 2002-no. 20.1. They give the external costs for several energy sources: lignite 3.8, coal 2.6, gas 1.1, photovoltaic 0.8, nuclear 0.2, wind 0.09 and hydro 0.02, all in eurocents/kWh. These figures are highly subjective and rest on questionable assumptions, such as the linear no threshold dose hypothesis for nuclear power. Nevertheless they are useful in drawing attention to the external costs as well as providing an incentive to obtain more accurate figures (Nuclear Issues 24, July 2002).

A useful parameter for evaluating energy sources is the ratio of the total energy output to total energy used to construct, maintain and decommission it. The average values of some estimates made by the International Atomic Energy Agency in 1994 are given in Table 3.5. The second column give the ratio of the total energy output to the total energy input and the third column is the reciprocal of this expressed as the percentage of input relative to output. It is notable that for some sources a wide range of values has been found.

Table 3.5. Energy Ratios for Various Sources (IAEA 1994)

Source	Energy Ratio	Input as Percentage of Output
Hydro	50	2
Nuclear	50	2
Coal	20	5
Gas	4–26	20–4
Solar	10	10
Wind	6–34	17–3

Table 3.6. Payback Times for Energy-Saving Measures (Royal Institution of Chartered Surveyors 2008)

Measure	Payback Time (years)
Cavity Wall Insulation	5
Double Glazing	124
Condensing Boiler	38
Solar Heating	208
Loft Insulation	13

Another useful measure is the payback time, that is the time taken for the device or energy-saving measure to produce the energy needed to construct and maintain it. The payback times for various energy-saving measures are given in Table 3.6.

3.11. Conclusion

This brief survey shows that at present the only energy sources apart from nuclear that are able to provide energy in the vast quantities needed are the fossil fuels oil, natural gas and coal. Other sources such as hydro remain useful, and as far as we can see at present the renewable sources are unable to provide more than a small fraction of what is needed. It has been estimated that if we relied on renewable sources alone, Britain could only support two million people, about thirty times less than the present population. Nevertheless, it is certainly important to keep an open mind about them, and to encourage further research and trials of any promising ideas. If any renewable source is shown to satisfy the requirements of capacity, reliability, cost, safety and respect for the environment it should be deployed on a large scale without delay. Until then, we have no alternative but to rely on existing energy sources. These are the fossil fuels which are responsible for global warming and serious pollution of the atmosphere and the earth. If we were to insist that we obtain all our energy from renewable sources, it can only be done by reducing our energy demands, which in turn means drastically reducing the population and our lifestyle. It is not easy to see how this could be done. If the renewables were the only possibilities, the outlook would indeed be grim, but during the twentieth century nuclear power has been developed, and this is considered in the next chapter.

References

Archer, Mary (2001) *UK Energy Policy Study: Emerging Technologies* (British Energy Position Paper).

Avery, John Scales (2007) *Energy, Resources and the Long-Term Future* (London: World Scientific).

Barnham, K.W., Mazzer, M., and Clive, B. (2006) Resolving the energy crisis — Nuclear or photovoltaics, *Nature* (Materials) **5**, 161.

Crabtree, George, and Lewis, Nathan (2007) In *Bright Future for Solar Cells* by Edwin Cartlidge, *Physics World* (July), p. 20.

Eikerling, M., Kornyshev, A., and Kucernet, A. (2007) *Driving the hydrogen economy. Physics World* (July), p. 32.

Evans, David (2004) Report. SONE No. 71.

Heath, Tom (2005) Wave energy comes of age, *Physics World* (May) p. 31.

Mabro, R. (ed) (1980) *World Energy: Issues and Problems* (Oxford: Oxford University Press).

Ross, D. (1995) *Power from the Waves* (Oxford: Oxford University Press).

Von Weizacker, E., Lovins, A.B., and Lovins, L.H. (1996) *Factor of Four: Doubling wealth — Halving Resources*, The New Report to the Club of Rome (Earthscan Publications Ltd).

Chapter 4

Nuclear Power

4.1. Energy from the Atom

In the previous chapter evidence has been presented indicating that the familiar power sources are either highly polluting or quite unable to provide the huge amounts of power that are needed to sustain our standard of living and also to raise that of people in the developing countries. There is however another energy source, the nucleus of the atom. This has been made possible by researches starting with the discovery of radioactivity by Bequerel in 1896 that eventually led to the discovery of the nucleus by Rutherford in 1911. In the following years there were many studies of different nuclei and their interactions.

Rutherford used the radiations from radioactive materials to bombard a range of nuclei, and he established that in some circumstances nuclear reactions can occur. In a nuclear reaction the incident particle can break up another nucleus when it collides with it or it can lead to the transfer of one or more particles from the projectile to the target nucleus, or from the target nucleus to the projectile. In nuclear reactions the sum of the masses of the interacting particles is usually different from the sum of those after the reaction. If it is greater, the excess mass is transformed into energy, which appears as kinetic energy of the emerging particles. The amount of energy released in one reaction is very small, so it was at first not considered practicable to extract useable energy from nuclear reactions.

The radiations from radioactive materials are very unsatisfactory tools for the study of nuclear reactions. They are rather low in energy and so are unable to overcome the electrostatic repulsion of the heavier target nuclei. They are not well-collimated and their energy cannot be continuously varied. This led to the development of nuclear accelerators that produced well-collimated monoenergetic beams that made possible the determination of the cross-sections and other features of a wide range of nuclear reactions. Lawrence built the first cyclotron in the USA, and Cockcroft and Walton made an electrostatic accelerator in Cambridge. This machine could accelerate protons to 700 KeV and they used them to break up the nucleus of lithium. This was the first time an atom had been split using a man-made accelerator.

Neutrons, discovered by Chadwick in 1932, are electrically neutral, and so they are not repelled by the electrostatic field of the nucleus. They therefore easily enter a nucleus, increasing its mass by one unit. Fermi in Italy bombarded a whole series of nuclei with neutrons and discovered that many new nuclei are produced. Two German chemists, Hahn and Strassmann, found that among the nuclei produced when uranium was used was one that seemed to have the same properties as barium. Meitner and Frisch realised that the nuclei of uranium had separated into two fragments, one of which was barium. They verified this explanation by calculating the energy released as the fragments recoil due to electrostatic repulsion and found it to be the same as would be expected from the masses of the participating nuclei. Making use of a biological term applied to the division of a cell into two they called the process fission. During the fission process two or three neutrons are also emitted, and these can enter nearby uranium nuclei and cause them to fission as well. This immediately raises the possibility of a chain reaction that results in the fission of a large number of the uranium nuclei. This happens very rapidly and causes a violent explosion and is the basis of the atomic bomb.

Uranium occurs naturally in the form of two isotopes, with the same nuclear charge but different numbers of nucleons, 235 and 238 respectively. It is only the uranium 235 that undergoes fission in this way, and it constitutes only 0.7% of natural uranium. In natural uranium so many neutrons are captured by the uranium 238 without causing fission that the chain reaction cannot take place. To make a bomb it is necessary to separate the uranium 235 from the uranium 239, a very difficult process that was achieved in the USA, thus making the bomb possible.

4.2 Nuclear Reactors

It is also possible to achieve a nuclear chain reaction in a controlled way with natural uranium. The difficulty is that so many of the neutrons are captured by the uranium 238 without causing fission that the chain reaction cannot take place. Fermi realised that if the neutrons from the fission can be slowed down away from the uranium 238 and then allowed to interact with the uranium 235 they would be able to cause further fissions and so sustain a chain reaction. The reason for this is that slow neutrons of about 0.1 eV are much more likely to cause fission in uranium 235 than to be captured by uranium 238. This can be done by forming the uranium into rods surrounded by a material called a moderator that slows the neutrons down by elastic collisions. The moderator must be as light as possible and must not capture neutrons, and the most convenient material with these properties was carbon. A reactor using this idea was first made by Fermi in 1942 in Chicago. The energy released during the fission process is mostly in the form of kinetic energy of the fission fragments. Since the time taken by a neutron to travel from one nucleus to the next is very small, a chain reaction produces a large amount of energy in a very short time.

The reactor is controlled by rods made of a material such as cadmium that absorbs neutrons and can be pulled in and out of the reactor. To start up the reactor, the rods are gradually pulled out until the multiplication factor is one, meaning that on average one of the neutrons emitted from each fission causes another fission. To increase the power of the reactor the rods are pulled out until the power rises to the required level and then re-inserted to bring the multiplication factor back to one. The time between one fission and the next is about a thousandth of a second, so the reaction takes place extremely rapidly and would be very difficult to control if it were not for the very convenient feature that some of the fission fragments emit a neutron some time later. If the multiplication factor is then set just below one, so that it needs the delayed neutrons, as they are called, to reach unity, it then becomes easy to control the reactor. If the multiplication factor exceeds one without the delayed neutrons the power rises extremely rapidly and the reactor becomes uncontrollable. This situation is called prompt critical, and is what happened at Chernobyl.

It was immediately realised that the heat generated by the fission process could be used as a source of power. To do this, water is circulated through the reactor in pipes and thence to a turbine where it can generate electricity in the usual way. A notable feature of nuclear reactors is that 1 kg of uranium 235 can produce 60 TJ of energy and this, assuming 30% efficiency of conversion, gives 3 GWh of electricity. Each fission releases 200 MeV compared with the few eV produced by the combustion of a carbon atom. Thus a 1 cm reactor fuel pellet produces the same amount of energy as 1.5 tonnes of coal. Further details of fission and the design and functioning of nuclear reactors may be found in many books (Hodgson 1999; Hodgson, Gadioli and Gadioli Erba 1997).

The atomic bombs brought the war with Japan to an abrupt end, and soon after several countries began to design and build nuclear reactors to produce electricity. Initially there was much optimism concerning the potentialities of the new energy source (Weinberg 1995). Nuclear power already contributes about 17% of the world's electricity, or 5% of world energy consumption. It is unpopular in some respects, and the rate of construction of nuclear power stations has fallen sharply in recent decades. However, there is now growing concern at the pollution of the atmosphere and the likelihood of climate change due to the reliance on fossil fuels. It is therefore urgently necessary to examine critically whether nuclear power should be abandoned or whether it can provide the energy we need.

To evaluate the present state of nuclear power, it is necessary to examine its capacity, cost, reliability, safety and effects on the environment. The cost of nuclear power has been discussed in Section 3.10 and its safety and effects on the environment will be considered in the next chapter. In this chapter the capacity and reliability of nuclear power are discussed in Sections 4.3 and 4.4, nuclear radiations in Section 4.5 and nuclear waste in Section 4.6. Future nuclear reactors are reviewed in Section 4.7 and fusion power in Section 4.8.

4.3. The Capacity of Nuclear Power

It is already clear from the figures presented in Table 1.1 that nuclear reactors are a major energy source. Nuclear reactors have been built in many countries and their energy production is given in Table 4.1. Nuclear accounts for about 80% of the electricity generated in France, and smaller but still substantial amounts in other countries, mainly in Europe, where nuclear energy production now exceeds that of coal. Uranium is plentiful and since its cost is a small fraction (15–20%) of the total it will still be economic even if poorer deposits than those now used have to be mined. Even if it does become uneconomic to mine, it will be possible to burn the uranium 238 in fast reactors, as discussed in Section 4.7. Since uranium 238 is about 140 times more plentiful than uranium 235 such reactors can operate for the foreseeable future. Present estimates are that the supplies of uranium will last about 300 years. Nuclear power stations cost £1100 to 1400 per kW to build, similar to the cost of coal power stations.

France, now producing about 80% of its electricity by nuclear power, exports nuclear-generated electricity to several surrounding countries. In 2001 this amounted to 15.8 TWh to Italy, 15.2 to Germany, 14.7 to the UK, 8.3 to Belgium and 7.4 to Switzerland (Nuclear Issues, December 2001).

The capacity of nuclear power can be illustrated by comparing the amounts of other sources required to equal the output of a 1000 MWe

Table 4.1.   The Capacity of Nuclear Power (Nuclear Issues 23, May 2001).

Country	Reactors in operation		Reactors under construction		Electricity supplied in 2000 (1999)		Operating experience		Load factor
	No of units	Capacity MW(e)	No of units	Capacity MW(e)	TWh(e)	Percentage of total	Yrs	Mths	Percentage
Argentina	2	935	1	692	5.7 (6.6)	7.3 (9.0)	44	7	69.6
Armenia	1	376			1.8 (2.1)	33.0 (36.4)	33	3	54.6
Belgium	7	5712			45.4 (46.6)	56.8 (57.7)	170	7	90.7
Brazil	2	1855			5.6 (4.0)	1.5 (1.3)	19	3	34.5
Bulgaria	6	3538			18.2 (14.5)	45.0 (47.1)	113	2	58.7
Canada	14	9998			68.7 (70.4)	11.8 (12.7)	433	2	78.4
China	3	2167	8	6420	16.0 (14.1)	1.2 (1.2)	23	5	84.3
Czech Rep.	5	2569	1	912	13.6 (13.4)	18.5 (20.8)	58	9	60.4
Finland	4	2656			21.1 (22.1)	32.2 (33.1)	87	4	90.7
France	59	63073			395.0 (375.0)	76.4 (75.0)	1169	2	71.5
Germany	19	21122			159.6 (160.4)	30.6 (31.2)	591	1	86.3
Hungary	4	1729			14.7 (14.1)	42.2 (38.3)	62	2	97.1
India	14	2503	2	606	14.2 (11.5)	3.1 (2.7)	181	5	64.8
Iran			2	2111					
Japan	53	43491	3	3190	304.9 (306.9)	33.8 (36.0)	962	8	80.0
Lithuania	2	2370			8.4 (9.9)	73.7 (73.1)	30	6	40.5
Mexico	2	1360			7.9 (9.6)	3.9 (5.0)	17	11	66.3
Netherlands	1	449			3.7 (3.4)	4.0 (4.0)	56	0	94.1

(*Continued*)

Table 4.1. (*Continued*)

Country	Reactors in operation		Reactors under construction		Electricity supplied in 2000 (1999)		Operating experience		Load factor
	No of units	Capacity MW(e)	No of units	Capacity MW(e)	TWh(e)	Percentage of total	Yrs	Mths	Percentage
Pakistan	2	425			1.1 (0.7)	1.7 (1.2)	29	10	
Romania	1	650	1	650	5.1 (4.8)	10.1 (10.7)	4	6	89.6
Russia	29	19843	3	2825	119.7 (110.9)	15.0 (14.4)	671	6	68.9
South Africa	2	1800			13.0 (13.5)	6.6 (7.4)	32	3	82.5
South Korea	16	12990	4	3820	103.5 (97.8)	40.7 (42.8)	169	2	91.0
Slovak Rep.	6	2408	2	776	16.5 (13.1)	53.4 (47.0)	85	0	78.2
Slovenia	1	676			4.5 (4.5)	37.4 (36.2)	19	3	76.0
Spain	9	7512			59.3 (56.5)	27.6 (30.1)	192	2	90.1
Sweden	11	9432			54.8 (70.1)	39.0 (46.8)	278	1	66.3
Switzerland	5	3192			25.0 (23.5)	38.2 (36.0)	128	10	80.4
Taiwan	6	4884	2	2560	37.0 (36.9)	23.6 (25.3)	110	1	86.5
UK	35	12968			78.3 (91.2)	21.9 (28.9)	1203	4	68.9
Ukraine	13	11207	4	3800	72.4 (67.4)	47.3 (43.8)	240	10	73.7
USA	104	97411			753.9 (719.4)	19.8 (19.5)	2455	8	88.3
Totals	438	351327	31	27756	2447.5 (2394.6)		9819	11	

nuclear reactor in one year. These are 6000 wind turbines; the burning of 30,000 km squared of forest; 2.3 million tons of coal; 1.9 million tons of oil; 18,000 cubic metres of gas or 100 square km of solar panels. For any given country, these figures have to be multiplied by the number of power stations needed.

In the next forty years about 2000 fossil fuel power stations must be replaced. Can this be done by building 4000 windmills on 500 sq.km each week? Or by covering 10 sq.km of desert each week with solar panels? Or by building 50 nuclear power stations each year? Such a rate of building is quite practicable as 43 were built in 1983 alone. There is thus no doubt that nuclear power can be expanded to meet the major part of our energy needs.

It is therefore quite practicable to bring the share of generation of nuclear electricity in the UK up to 80% by the year 2020. This is confirmed by the French achievement; they increased their output from around 50 TWh in 1980 to 400 TWh in 2000. At one time they were commissioning a new reactor every two weeks (Nuclear Issues, March 2008).

4.4. Reliability

Nuclear reactors are extremely reliable; and many of the world's reactors operate over 90% of the time, and the remainder of the time is mostly devoted to essential maintenance, which can be scheduled well in advance to take place when the demand is low. Unplanned breakdowns are increasingly rare. The unit capability factor, defined as the percentage of maximum energy generation that the plant is capable of supplying to the grid, is continually increasing. The number of unplanned shutdowns has fallen to about one in 7000 per day (Nuclear Issues 22, April 2000).

A disadvantage of nuclear reactors is that they take longer to come up to full power than other power stations, especially those using gas. Nuclear power stations are therefore best to provide the base load, with alternative sources such as gas power stations that can be rapidly activated when there is a sudden need.

4.5. Nuclear Radiations

One of the main differences between nuclear and other power stations is the presence of nuclear radiations. The fission fragments produced when the uranium nuclei split are highly radioactive and emit alpha-particles, beta and gamma rays until finally a stable nucleus is formed. There are many different nuclei among the fission fragments, and their half-lives vary from a small fraction of a second to many thousands of years. These decay rates are characterised by a half-life, which is the time taken by the radioactivity of a sample of a particular type of nucleus to decay to half its initial value.

When they pass through the human body nuclear radiations can break up the complicated molecules inside the cells, releasing reactive radicals that can cause more damage. If the level of radiation is small few cells are affected; they are soon replaced and no harm is done. If, however, the radiation level is high, serious damage will be caused and cancers may develop during the following years. In the case of massive whole-body irradiation death can also take place. It is vital, of course, to specify just what we mean by low and high levels of irradiation, and this will be done later.

The three types of nuclear radiation have different effects on the human body. Alpha-particles are nuclei of helium and since they are doubly charged they lose energy rapidly and ionise strongly, and are very destructive. Their short range means that they are harmful only if the radioactive material is inside the body. The beta rays are energetic electrons and the gamma rays are short wavelength electromagnetic radiation. They can both penetrate far inside the human body.

Nuclear radiations can easily be detected by very sensitive instruments that can record the passage of a single particle, so it is possible to detect the presence of extremely small amounts of radioactive substances. This enables us to learn how they move through the atmosphere, the oceans and our own bodies. This property has proved to be extremely useful in medical research.

When considering the effects of nuclear radiations on people, it is necessary to take account of the different sensitivities of the different organs of the body. This is done by defining the rem, which is the dose given by gamma radiation that transfers a hundred ergs of energy to each gram of biological tissue, and for other types of radiation it is the amount that does the same biological damage. A new unit, the Sievert, has now been defined as 100 rem.

Nuclear radiations are often feared because they are unfamiliar and can cause great damage to living organisms without our being aware that anything untoward is happening. The damage only appears afterwards, sometimes very long afterwards, when it is too late to do anything about it. Our senses warn us of many dangers, such as excessive heat and some poisonous gases, and we can take avoiding action. Nuclear radiations are not alone in being invisible; many poisonous gases such as carbon monoxide have no smell, and we do not know that a wire is live until we touch it and receive an electric shock.

When nuclear radiations were first discovered, they were welcomed with enthusiasm, and to some extent this was justified. In the form of X-rays they improved medical diagnosis and treatment, and bottles of health-giving mineral waters were advertised as radioactive. It was only much later, when pictures were released of the radiation damage to the victims of Hiroshima and Nagasaki, that the public image of nuclear radiations switched to one of fear.

Undoubtedly this reaction has gone too far. Nuclear radiations are indeed dangerous in large amounts, but so are fire and electricity. Properly used nuclear radiations have numerous beneficial applications in medicine, agriculture and industry. Like so many of God's gifts, they can be used for good or evil.

Nuclear radiations are not new; they did not first enter the world with the experiments of Henri Becquerel or Madame Curie. They have been on the earth since the very beginning. Many rocks and minerals, such as the pitchblende refined by Madame Curie to produce the first samples of radium, are naturally radioactive and emit nuclear radiations all the time. The nuclei formed by such radioactivity include radon, a gas that seeps up through the soil and enters our homes. The natural radioactivity of the earth varies greatly from one place to another, depending on the concentration of rocks containing uranium as shown by Table 4.1. In addition, the earth is bathed in the cosmic radiation from outer space, and they are passing through our bodies all the time. Cosmic rays are attenuated as they pass through the atmosphere and so they are more intense at the top of a mountain than at sea level. There are radioactive materials in our own bodies, such as a rare isotope of potassium. Thus the human species has evolved through millions of years immersed in nuclear radiations. This natural radioactivity is important for estimating the hazards of nuclear radiations in general, since if the additional source emits radiations at a level far below that of the natural radiation it is unlikely to be injurious to health.

In addition to this natural radiation, we are exposed to radiation from medical diagnosis using X-rays, medical treatment, atomic bomb tests and the nuclear industry. Estimates of the radiation exposure in the United Kingdom due to all these sources are given in Table 4.1. This for medical purposes is quite high but in the long term what is important is the average exposure over a long time weighted by the age distribution of those exposed. This is because the effects of radiation at levels typical of medical uses do not appear for many years so that the irradiation of young people before the end of their reproductive age is more serious than that given to older people. Since the larger part of the medical irradiation is received during the treatment of cancers, which more often afflict older than young people, the dangers to health due to medical irradiation are not so great as might appear.

As shown in Table 4.2, nearly half the radiation exposure due to the natural background is attributable to radon. This is a radioactive gas formed by the radioactive decay of uranium. In regions where the soil contains uranium the radon seeps upwards into the atmosphere or into our homes where it collects unless the house is well ventilated. Radon decays with the emission of alpha-particles and when breathed in can irradiate the inside of the lung, causing lung cancer. According to the National Radiation Protection Board a radon gas concentration level of 200 Bq/m^3, equivalent to an effective dose of 10 mSv per year, is the level at which action should be taken to reduce the level. This involves creating a cavity under the floors and pumping out the radon at a cost of up to £1000. Many local authorities are now recommending that such action

Table 4.2.   Radiation Exposure in the United Kingdom.

Source		Exposure in Millirem per year		%
Cosmic Radiation		31		12
Terrestrial gamma rays	Natural	38	186	13.5
Internal Radiation		37		10
Radon Decay Products		80		50
Medical Irradiation		50		14
Fallout from bomb tests		1		0.2
Miscellaneous sources	Man-made	0.8	53	0.3
Occupational Exposure		0.9		0.3
Disposal of Radioactive Wastes		0.3		0.1

(*National Radiological Protection Board Bulletin*, March 1981; *Atom*, May 1981). The percentage contributions are from a later study (SONE 2006).

be taken. However, before doing this, it is necessary to establish the relation between the level of exposure and the probability of lung cancer. Many studies world wide, in Canada, China, Finland, France, Germany, Japan, Sweden and the USA, have failed to establish any positive correlation and indeed in three of these studies there was an inverse relationship. Other studies (Darby *et al.* 2004) find that the increased risk of lung cancer due to a lifetime dose due to 100 Bq/m is about 0.1% and 25 times greater for smokers. The data used in this study were consistent with a linear dose relationship, but do not exclude different behaviour at very low energies. The validity of this assumption is discussed in more detail below. It thus seems that, particularly for non-smokers, the level of irradiation due to radon is so low that when compared with other much greater hazards it is difficult to justify such expensive precautions.

Radioactive isotopes have many medical applications. If, for example, we want to know how salt is taken up by the body, we can feed a patient with some salt that contains a very small amount of a radioactive isotope of sodium. This emits radiation that can be detected by a counter outside the body, and so we can follow the progress of the sodium as it is absorbed. The amount of radiosodium needed is so small that it does no harm to the patient. In this way radioisotopes provide a valuable diagnostic tool. Radioisotopes can also be used for treatment. Thus it is known that iodine tends to concentrate in the thyroid gland. If therefore we want to treat cancer of the thyroid we can feed the patient with radioiodine, and it will go to the thyroid gland and irradiate the tumour, without appreciably affecting the rest of the body. Much can be learned from experiments on animals, but it is always hazardous to extrapolate from mice to men.

The powerful nuclear accelerators that are used to explore the structure of the nucleus and to produce new unstable particles can also be used to irradiate tumours. The radiation emitted by radium and other natural sources has the disadvantage that it is relatively low in energy and so can penetrate only a small distance into the body. In addition, the radiation comes out in all directions equally. If we want to treat a tumour deep inside the body we need a way of irradiating the tumour that minimises the irradiation of the surrounding healthy tissue. The only way to do this is to have a collimated beam of radiation of sufficient energy to penetrate the body, and such beams are produced by accelerators. During the treatment, the patient is rotated so that the beam always passes through the tumour but irradiates a particular part of the surrounding healthy tissue for only a small part of the time. This is a difficult technique, but with great care it can be used successfully. Many nuclear accelerators such as that at Faure in South Africa are used partly for medical treatment and partly for nuclear research.

There are other techniques that can be used to see inside the body for diagnostic purpose. Nuclear magnetic resonance and ultrasound are routinely used to provide a detailed picture of internal organs. At a very early stage in pregnancy it is possible to examine a baby in the womb to check that all is well. Long thin fibre optic probes can be inserted far into the body to take pictures of diseased organs, providing valuable data for the surgeon before he operates. Microsurgical techniques allow operations to be performed with minimum disturbance to healthy tissue. Many operations that formerly required large incisions and hence protracted convalescence can now be done so easily that the patient can leave the hospital the same day.

Sometimes it is difficult to know whether the benefits of radiation outweigh the hazards. Thus X-rays can detect cancers early enough for effective treatment, and yet they can also themselves cause cancers. A detailed study of stomach tumours showed that for young people the dangers outweigh the benefits, whereas for older people the opposite is the case.

The main radioactive hazard from a nuclear reactor comes from the fission fragments. These are intensely radioactive and some of them continue to emit radiations after thousands of years. In normal operation, nuclear power stations emit very small amounts of radioactivity into the atmosphere. These amounts are even less than the radioactive emissions from coal power stations. Professor Fremlin has estimated that the radiation dose received in Britain due to the nuclear power stations reduces our average life expectancy by one or two seconds.

There is widespread public anxiety about the effects of nuclear radiations, particularly concerning the genetic effects and the cases of leukaemia in children near nuclear installations. The children of the survivors of the atomic bombing of Hiroshima and Nagasaki, who all received massive doses of radiation, have been studied in detail by Professor S. Kondo, who personally visited Nagasaki soon after the bombing and saw the devastation. He has studied the effects of the bombing for forty years and has recorded the indicators of genetic damage for 20,000 children of atomic bomb survivors exposed to an average dose

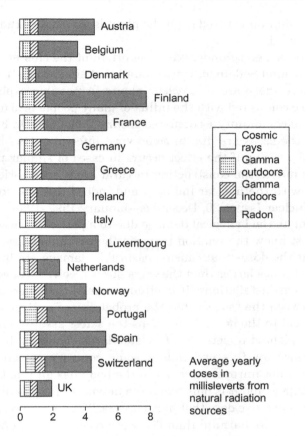

Figure 4.1. The Average Yearly Doses from the Natural Radiation Sources in European Countries (National Radiological Protection Board (NRPB), 1993).

of 400 mS. The numbers of the genetic indicators chromosome abnormalities, mutations of blood proteins, childhood leukaemia, congenital defects, stillbirths and childhood deaths showed no differences between the children of the atomic bomb survivors and a control group. There is thus no evidence of genetic damage due to the atomic bombs.

Seven cases of childhood leukaemia occurred between 1955 and 1983 in Seascale in Cumbria near the nuclear reprocessing plant at Sellafield. This number seemed to be greater than would be expected by chance, and it received much publicity. It was, however, very difficult to see how these cases could be blamed on Sellafield, since the amount of radiation emitted from the plant is far smaller than the natural background.

It still remained to understand the origin of the cluster of cases of leukaemia around Sellafield. It was found that similar clusters occur in regions where there are no nuclear plants. A possible explanation is that they are connected with the influx of many people into a relatively isolated rural community, as occurred around Sellafield. Dr. Kinlen proposed that the cases are due to some viral infection. This hypothesis was tested by seeing if the effect occurs in cases of similar population movements due to the construction in rural areas of factories with no connection with the nuclear industry and indeed this was found to be the case (Nuclear Issues 20, December–January 1997–1998).

To estimate the biological damage due to a particular dose of radiation we must know the relation between the two quantities. The difficulty is that the doses that cause measurable damage are hundreds of thousands of times larger than the extra doses received by people living around nuclear installations. It is often assumed that there is a linear relation between the two, so that the probability of contracting cancer is proportional to the dose. As it seems the safest assumption to make it is widely adopted in setting safety standards. There is, however, no direct evidence for this and indeed some contrary evidence (Cohen 1999). This is not unreasonable, since the body has an innate capacity to repair damage and it is only when the defences of the body are overwhelmed by a massive dose that harm occurs. Thus a dose received over a long period is less harmful than if it were received all at once.

A direct result of the linear dose assumption is the setting of unreasonably strict limits on permitted radiation exposure in many industries, thus greatly increasing costs. This leads to reluctance to accept vital radiodiagnostic and radiotherapeutic irradiations, and restricts the use of radiation in industry and research. Adherence to these exposure limits led to large-scale evacuation from the region around Chernobyl, causing much unnecessary distress and suffering. It has been estimated that the present regulations imply an expenditure of $US 2.5 billion to save one hypothetical life whereas it costs $50 to $100 to save a life by immunisation against measles and diphtheria in the developing countries (Nuclear Issues 22, August 2000).

The linear dose assumption is critically important when estimating the effects of small doses from a reactor accident. Thus it has led to an estimate of 9000 to 33,000 latent fatal cancers due to the Chernobyl accident. There is no direct evidence for this and it is unlikely because studies by UNSCEAR (the scientific committee of the United Nations responsible for studying the effects of atomic radiations) said in June 2000 that for the most heavily contaminated area, the Bryansk region, and apart from 1800 cases of curable cancer cases in children 'there is no evidence of a major public health impact attributable to radiation exposure fourteen years after the accident'. This conclusion was supported by four Russian scientists who confirmed 'that in all the years since the Chernobyl accident there has been no significant divergence in the overall mortality rate among the population of the contaminated area of Russia'. Furthermore, 'the overall mortality rate among the cleanup staff, who generally receive higher doses of radiation than the general public, was statistically lower than the mortality of the control group over all the years following the accident' (Nuclear Issues 23, March 2001).

It is also possible that small doses stimulate the body's repair mechanisms, so that small doses are beneficial (Fremlin 1987). This is supported by an extensive study made by Frigerio *et al.* at the Argonne National Laboratory in 1973. They compared the cancer statistics for the USA from 1950 to 1967 with the average natural background for each State, and found that the seven states with the highest natural background had the lowest cancer rates. Unless there is some other explanation for this result it implies that the chance of contracting cancer is reduced by 0.2% per rem. Further evidence is provided by 'the higher life expectancy among survivors of the Hiroshima and Nagasaki bombs; many times lower incidence of thyroid cancer among children under fifteen exposed to fallout from Chernobyl than the normal incidence among Finnish children; and a 68% below average death rate from leukaemia among Canadian nuclear energy workers' (Taverne 2004). Many studies on animals have given similar results. Furthermore, it is found that people living in areas of high background radiation show no evidence of detrimental effects; thus a study by the Regional Cancer Centre at Trivandrum in Kerala of 300,000 people living in an area with a background dose of 200 mSv/y found a life expectancy of 74 years compared with 54 years for India as a whole. This dose may be compared with the permitted occupational levels laid down by the ICRP of 20 mSv/y and the population dose of 1 mSv/y. No carcinogenic or leukaemogenic effects have been observed for radiation doses less than 200 mSv/y. An excess of leukaemias has been found only for doses more than 400 mSv and for tumours of 1000 mSv (Nuclear Issues 20, June 1998 and 23, April 2001). Aircrews are exposed to higher doses of cosmic radiation, and so their union asked for compensation. Studies of the mortality rates of 19,184 pilots in the period 1960–1996 showed however that they actually decreased with increasing dose. The skin cancer rate was however higher because of the time they spent lying in the sun on tropical beaches. Such evidence has been widely discounted because it seems counter-intuitive.

Additional evidence is supplied by the observation that 'Hiroshima and Nagasaki survivors with low radiation doses are outliving those who received no dose at all. Radiation-exposed US navy shipyard workers of the 1960s and 1970s have lower cancer rates than those not exposed. Canadian women who received small doses of radiation in hospital had breast cancer rates two-thirds those of others' (Speakers' Corner 96, May 1997).

In favour of the idea of a threshold dose, it can be argued that the passage of a single nuclear particle through a cell, the lowest possible dose, can cause DNA double strand lesions. Such lesions occur naturally at the rate of about ten thousand per cell per day, whereas exposure to radiation at the current population exposure limit would cause only two lesions per cell per day. Thus radiation-induced lesions are insignificant compared with those occurring naturally. This implies that the 'risks of health effects are either too small to be observed or are non-existent at doses below a threshold value of 150 mSV or even up to 500 mSV' (Nuclear Issues 25, February 2003).

A new technique for evaluating the effects of small doses of radiation has been developed by Professor Feinendegen. His results show conclusively that the linear dose assumption is incorrect: at low doses there is an additional quadratic term. Furthermore, a Joint Report of the Academie des Sciences (Paris) and of the Academie National de Medicine concludes that estimates of the carcinogenic effects of low doses of ionising radiations obtained using the linear assumption could greatly overestimate those risks (Tubiana and Aurengo, Nuclear Issues, October 2005).

An extensive study by Darby *et al.* (2005) of the risk of lung cancer from radon found a linear dependence on the radioactivity of the air from zero to 1200 Bq/m^3. The data however are not inconsistent with a threshold for doses up to 150 Bq/m^3. Radon is particularly dangerous because it is a gas that is breathed directly into the lungs. Other sources, such as the cosmic radiation, consist of singly charged particles that pass through the body and thus give it a whole-body dose. Since it is likely that doubly-charged particles are much more effective in breaking the DNA than singly-charged particles it is unjustified to assume that the results of the radon study apply to all radiations. In the case of the exposure due to Chernobyl, it is possible that some radioactive dust particles will reach the lungs, but most will be stopped before they get there. The radon results thus provide no convincing evidence against the existence of a threshold dose.

Another possible mechanism to account for the Sellafield cluster was suggested in 1987 by Gardner, who postulated that the children developed leukaemia as a result of their fathers' exposure to nuclear radiation. He collected statistics that showed a significant correlation between paternal radiation dose and leukaemic children. This led to several court cases by families seeking compensation from British Nuclear Fuels, the company operating the plant.

The Gardner hypothesis has such serious implications for the nuclear industry that many further studies were made. These included the actual process whereby paternal irradiation could lead to childhood leukaemia, the statistics of leukaemia in the children of survivors of the atomic bombing of Japan, and more extensive studies of leukaemia around nuclear plants. The results of these studies were published by Sir Richard Doll, Dr. H. J. Evans and Dr. S. C. Darby (1994).

They showed that the possibility that nuclear irradiation could cause a gonadal mutation leading to childhood leukaemia can be studied using data on genetically-determined leukaemia. The detailed statistics shows that there may be a recessive mutation that could contribute to a number of the observed cases. However, 'it effectively excludes and major contribution from the type of mutation that would be required to account for the appearance of the Sellafield cases in the first generation, namely a dominant mutation with a high degree of penetrance'.

Another study of over 20,000 children of employees of British Nuclear Fuels and the UK Atomic Energy Authority showed that these children were no more likely than any other in the general population to develop any kind of cancer, including leukaemia. Further studies showed 'no link between birth defects, miscarriages, or stillbirths and paternal pre-conception irradiation'. The same applies to the very small number of women working in radiation areas during pregnancies (Nuclear Issues 22, October 2000).

Studies by Neel and colleagues of 'the children of atomic bomb survivors including more that 1500 born to parents who received a gonadal dose of one sievert or larger, revealed no clearly increased frequency of mutations'. These doses are far higher than those received by the Sellafield workers.

Further studies were made of all the leukaemia cases in people under 25 years of age, born after 1958 and between the years 1958 and 1990, in Scotland and a part of north Cumbria near the Scottish border, and of all children under 15 born near five nuclear installations in Ontario. They found that 'neither set of results supported the probability of a hazard from the father's occupation'. Several other studies reached the same conclusion.

Thus the authors conclude that 'the association between paternal irradiation and leukaemia is largely or wholly a chance finding'. They note that there appear to be 'small but real clusters of leukaemia in young people near Sellafield, and some other explanation for them needs to be sought'.

This highly authoritative study should finally lay to rest the fears of radiations from nuclear installations, but whether it will or not depends on the mass media. The presence of leukaemia clusters, and particularly the Gardner hypothesis, has been widely publicised by organisations opposed to nuclear power. This has encouraged families with children suffering from leukaemia to seek compensation, but when the scientific evidence is laid before a court, the judgement inevitably goes against them.

There is also some concern about the radiation dose received by people who eat sea food from the Sellfield region. Studies have shown that the few people who eat very large amounts may receive an extra annual dose of 0.35 mS. Those living near Sellafield may receive an extra dose of 0.25 mS. This is to be compared with the average annual background dose of 2.2 mS per year and about 8 mS in Cornwall. Similar studies on other countries give the same results.

The Danish Minister for the Environment has criticised Sellafield for releasing the isotope technicium 99 into the sea. Measurements in the Kattegat show that the resulting radiation level is between two and three Becquerels (Bq) per cubic metre and 0.1 Bq/kg in fish and 20–25 Bq/kg in lobsters. This may be compared with the natural radioactivity of sea water that amounts to 12,000 Bq/m^3. If a fish-lover consumes 50 kg of fish and 20 kg of shellfish per year, the resulting radiation dose is about 0.14 microSv. For comparison, a person inside a typical Danish house receives an annual dose of about 30 microSv so that while eating the fish the dose from the air is about 200 times that from the fish. Furthermore, all fish and shellfish contain polonium 210, and this gives a dose about 300 times that from the technicium (Nuclear Issues 22, May 2000). The technicium discharges from Sellafield have now ceased.

It has been claimed that the Irish Sea is the most radioactively contaminated in the world. In fact, measurements show that the eastern Mediterranean, the Persian Gulf and the Red Sea, where thousands of tourists swim every year, are ten times more radioactive. The Irish Radiation Protection Board has now estimated that the radiation dose from the Sellafield discharges to a heavy fish eater is 0.00133 mS/y and for a typical consumer 0.0003 mS/y (Nuclear Issues 24, September 2002).

A comparison with other sources of radioactivity shows that 'the major source of man-made radioactivity comes from the North Sea oil and gas operations of Denmark, Norway and the UK', and this is much smaller than the natural sources. 'The collective radiation from all natural background sources for the European community is put at 844,000 mS/y. The collective dose from the natural isotopes present in sea water is bout 17,000 mS/y. In the year 2000, the total collective dose from all man-made discharges into the sea amounted to 300 mS/y, of which 260 mS/y came from the phosphate fertiliser and oil industries. The contribution from all radioactive isotopes discharged by the nuclear industry, including the fuel reprocessing plants at Sellafield and La Hague, was only a minute 14 mS/y, over 1000 times less than the natural activity in the seawater' (Nuclear Issues, April 2003).

Detailed studies by the National Radiation Protection Board and the International Commission on Radiation Protection of the life histories of thousands of workers have shown 'that there is no evidence that radiation workers have cancer mortalities which are higher than those for the general population'. If anything, the statistics showed smaller rates. A study of workers in the UKAEA in 1985 showed that their death rates from cancer were 22% lower than the national average, and 14,347 workers at Sellafield had an average mortality rate from all causes that was 2% less than the national average. A study of 21,358 men who took part in the UK atomic bomb tests in Australia and the Pacific showed no detectable effect on their life expectancy or on the incidence of cancer and other diseases.

The concern about nuclear radiation has diverted attention from other threats to our health. Radiation is responsible for only about 1% of diseases worldwide, and most of this comes from the natural background and from medical uses. The nuclear industry is responsible for less than 0.01%. The vast sums spent to reduce this still further could be spent far more effectively on simple disease prevention. It is greatly in the public interest that these matters should be treated as objectively as possible, taking full account of the scientific evidence. This would avoid much unnecessary anxiety and enable the best decisions to be taken concerning our future energy supplies.

4.6. Nuclear Waste

In 1976, a report by the Royal Commission on Environmental Pollution in the UK under the Chairmanship of Lord Flowers stated that: 'There should be no commitment to a large programme of nuclear fission power until it has been demonstrated beyond reasonable doubt that a method exists to ensure the safe containment of long-lived highly radioactive waste for the indefinite future'. Since that time the problem of nuclear waste management has loomed large in the public discussions on nuclear power and has been repeatedly emphasised by those opposed to it.

One of the main hazards of the operation of nuclear reactors comes from the waste that they produce, in particular the fission products that remain in the fuel rods. As the uranium or plutonium is burnt in the nuclear reactor, the fission products accumulate until they absorb so many neutrons that they prevent the reactor from working. To avoid this, spent fuel rods are continually removed from the reactor and replaced by new ones. The spent fuel rods are taken to the reprocessing plant where the uranium and plutonium are separated and used to make new fuel rods. The remaining portion contains the highly radioactive fission fragments. As already mentioned, the fission fragments are a mixture of many different nuclei with widely varying half lives. The first step in the disposal of the fission fragments is to store them in safe tanks above ground for a few decades so that most of the radioactivity from the short-lived nuclei decays away. Then the remainder is concentrated and fused to form a glassy or ceramic substance. For extra safety this is placed in stainless steel containers and then buried far below the surface in a stable geological formation such as clays, disused salt mines and granite. There is then no chance that the fission products will escape and cause harm. This has been checked by a detailed study sponsored by the European Community. Eventually, over the years, the radioactivity of the fission fragments will decay until it is similar to that of the surrounding rocks.

The amount of radioactive waste from nuclear reactors is not large. Every year, a nuclear power reactor produces about four cubic metres (cm) of high-level waste, 100 cm of intermediate-level waste and 530 cm of low-level waste. The total amount of high level waste produced in Britain from 1956 to 1986 is about 2000 cm, about the same volume as an average house. This is very small compared with the vast amounts of poisonous chemical waste produced by the manufacturing industries, much of which is buried in the sea or emitted into the atmosphere. The cost of waste disposal and decommissioning is estimated to be less than 0.001–0.002 eu/kWh.

Another estimate of the amount of waste from a nuclear power plant is given by Abel Gonzales in the International Atomic Energy Agency Bulletin. Every year it uses 27 tonnes of fuel and produces about 27 tonnes of high level radioactive waste, 310 tonnes of intermediate level waste and 460 tonnes of low level waste. These figures may be compared with the corresponding ones for a coal power station in Section 2.2.

The high-level nuclear waste consists of a mixture of uranium (95%), plutonium (1%) and the fission products (4%) that include long-lived radioactive elements. The uranium and plutonium can be separated by reprocessing and used as a reactor fuel. Thus most of the fuel elements are not waste but valuable material.

The stock of separated plutonium in Britain is about 135 tonnes, and there are also 8640 tonnes of depleted, natural and low enrichment uranium in the fuel supply. The plutonium and uranium are already recycled as reactor fuel, the plutonium as mixed oxide fuel and the uranium as new fuel. The energy content of one tonne of plutonium is equivalent to three million tonnes of coal, so that the value of the UK stock of plutonium is over £120M. Now that uranium costs up to $100 per pound, the value of the depleted, natural and low enrichment uranium is around £10B. There will be about 106,000 tonnes of depleted uranium in Sellafield by 2020. This has an energy content of 1000 MW thermal days/tonne, where one tonne is sufficient to fuel a 1000 MW reactor for a year. Taking the wholesale price of electricity to be 5p/kWh gives each tonne a worth of £600M, and so the total value is around £64T, enough to supply all UK homes, industry and electrical transport for 600 years. This uranium has the energy of 1500B barrels of oil equivalent, five times the total reserves of Saudi Arabia (Nuclear Issues 29, May 2007).

Medical and industrial processes using radioactive material also produce some low-level radioactive waste. This may be safely stored, together with similar waste from nuclear reactors, by burying it quite near the surface in a place that will not be disturbed.

To illustrate the quantities involved in nuclear waste disposal, a nuclear fuel pellet having about half the volume of a cigarette can generate enough energy to supply the needs of a family of four for a year. To obtain the same amount of energy from other sources would require a ton of oil, two tons of coal, five tons of wood or 30,000 cubic feet of gas. After reprocessing, the waste from the original pellet would be about the size of a pinhead. Newly-designed nuclear reactors will produce less than a tenth of the waste from existing ones.

A new type of reactor, suggested by Carlo Rubbia, is now being studied in CERN and Los Alamos. It is not only very safe, but also consumes its own waste. Ultimately it is hoped that it can be developed as a power reactor, but the first aim is to use it to destroy nuclear waste. The reactor is designed so that it is always sub-critical, so there is no possibility of an accident. To maintain the reaction more neutrons are required, and these are produced by spallation. To achieve this, a beam of high energy (about 1 GeV) protons is produced in a separate accelerator, and this beam hits the reactor, which is cooled by liquid lead. The protons collide with the lead and produce a shower of neutrons, about fifty for each incident proton. More neutrons are produced by fission, and the neutron flux is so high that the waste is consumed. This reactor is very safe because the reaction stops as soon as the proton beam is switched off. The reactor produces power in the usual way, and about 5 to 10% of it is sufficient to drive the proton accelerator. The fuel used in this reactor is a mixture of thorium and uranium 233 or plutonium 239 that are fissionable by slow neutrons. It is surrounded by pure thorium, chosen because it is plentiful and is transformed to uranium 233 by neutron capture. The reactor thus breeds and burns uranium 233. A proton beam of 15 mA can produce about 200 MWe; an associated advantage is that less plutonium is produced. The molten lead coolant transfers the heat of the reaction by convection to the heat exchangers. This type of reactor can burn actinides that are not fissionable by thermal neutrons. The energy produced in such a reactor by burning 1 kg of thorium is equivalent to that produced in a pressurised water reactor by burning about 250 kg of uranium.

Studies of ocean disposal have shown that the risks are vanishingly small. Nevertheless, the London Dumping Convention voted to impose a moratorium, providing yet another example of a decision imposed by political rather than scientific considerations. The opponents of nuclear power now reject not only waste disposal but also the scientific studies undertaken to find the best ways to do it. It may be mentioned that the seas already contain 4000 million tons of uranium and other radioactive elements, vastly greater than the amounts we want to put there.

The situation concerning the disposal of nuclear wastes has changed greatly since the report made in 1976. The view of the Chairman of the Commission, Lord Flowers, is that the conditions mentioned at the beginning of this section have now been met. 'He said that international efforts had demonstrated adequately the techniques of waste disposal and that it should no longer be used as an argument for holding up nuclear development' (Nuclear Issues 29, October 2007). According to the OECD Nuclear Energy Agency 'there is a broad scientific and technical consensus that deposition in deep geologic formations is an appropriate and safe means of isolating it from the biosphere for very long timescales' (Nuclear Issues, September 2000, p. 3).

At the end of its life, a nuclear reactor has to be decommissioned, that is taken out of service, dismantled and the site returned to its original state. The cost of this is met by setting aside each year a small proportion of the payments received from the sale of the electricity produced by the reactor. If 'decommissioning costs are included along with waste management the electricity still costs less than that from gas or coal or solar or wind' (Nuclear Issues 27, November 2005). Figures for the cost of decommissioning are given in Section 3.11.

4.7. New Nuclear Reactors

There are very many types of nuclear reactors, both actual and potential, and continual studies are in progress to design reactors that are safer and more efficient, reliable and economical. Reactors may be classified according to their purpose, their fuel and moderator and the energy of the neutrons when they cause another fission. A reactor may be built for experimental purposes, or to produce fissile material or to generate useful power. The fuel can be uranium, or uranium enriched with a higher proportion of fissile uranium isotopes, or with plutonium or thorium. As already mentioned, the neutrons in the reactor will cause fission only if they are slowed down, and this is done by including in the reactor a moderator. The moderator can be any material that slows the neutrons down efficiently by elastic collisions without capturing too many of them. The lighter the material the more efficiently it slows the neutrons, but many of the lighter nuclei are excluded because they readily capture neutrons. This leaves ordinary water, deuterium in the form of heavy water and graphite as the preferred choices. Reactors with a moderator are called thermal reactors. Fermi's first experimental reactor used natural uranium as fuel and graphite as a moderator because at that time heavy water was too expensive and ordinary water captures too many neutrons. Subsequently, large reactors were built to produce the fissile plutonium 239, which is formed from the non-fissile uranium 238 by neutron capture followed by the successive emission of two electrons. The plutonium differs chemically from uranium, and may therefore be separated from the uranium by relatively easy chemical methods.

It is not possible to build a reactor with natural uranium and ordinary water because the hydrogen nuclei in water combine with neutrons to form deuterons. Too many neutrons are absorbed in this way and the chain reaction cannot develop. If however the uranium is enriched to about 3% fissile uranium 235 or with plutonium 239 a chain reaction can occur with ordinary water as a moderator. These are the light water reactors that are widely used to generate nuclear power.

If the uranium is enriched to 10%, a moderator is not needed. Such reactors are called fast reactors and they can be designed to produce more fissile material than they consume, and hence are referred to as breeders. The main fuel is plutonium 239 and sufficient fissions occur to sustain the chain reaction and also to convert some of the uranium 238 into plutonium. One type of fast reactor contains 80% uranium 238 and 20% plutonium 239 in the interior, surrounded by a blanket of uranium 238. About three neutrons are emitted when plutonium 239 fissions, and of these one is required to sustain the chain reaction, leaving two to account for losses and to breed more plutonium 239. Another type of fast reactor uses a blanket of thorium 232, which is converted into the fissile uranium 233. The spent fuel rods from all these reactors are sent to a reprocessing plant such as the Thermal Oxide Reprocessing Plant (THORP) at Sellafield to extract the fissile material in the form of uranium-plutonium mixed oxide (MOX) reactor fuel. Japan has a large reprocessing plant at Rikkahao that is planned to reprocess about 430 tonnes of spent nuclear fuel to produce 2.3 tonnes of plutonium. This will be made into MOX fuel for use in their light water reactors (Nuclear Issues, January 2006).

The power density in a fast reactor is substantially higher than in a normal power reactor, so liquid sodium, which cools effectively without moderating the neutrons, is used to remove the heat. Prototype fast reactors have operated for a number of years in several countries. Their great advantage is that they enable over 50% of the uranium to be used to generate power instead of the 0.7% of natural uranium. They can either produce or burn plutonium, depending on the neutron density, and can also be used to burn nuclear wastes.

Fast reactors have a strong negative temperature coefficient and so are inherently safe. Due to the low level of corrosion and the ease with which components can be replaced they are likely to have a longer life than thermal reactors, possibly up to seventy years. They also conserve raw material and reduce the quantity of waste products.

Since fast reactors can burn the uranium 238 remaining in the spent fuel rods from thermal reactors, there is a vast store of energy waiting to be used. It is estimated that the depleted uranium now stored in Britain contains the energy equivalent of all the 500 years' supply now in our reserves. Fast reactors are also the most efficient way to use the plutonium and highly enriched uranium from the decommissioning of nuclear warheads. They cost more to build than thermal reactors, and so at present it is uneconomic to build them because uranium is plentiful. Ultimately it will become too costly to extract all the uranium we need for thermal reactors and then we can turn to fast reactors. It is thus unwise to abandon the development of fast reactors.

Nevertheless many countries, including the USA, Britain and France, have severely curtailed or cancelled their fast reactor programmes. Germany's fast reactor at Kalkar, costing £5B, was completed but never allowed to operate, a victim of politics. The 1200 MWe French fast reactor Super Phenix at Creys-Malville started to operate in 1985, but was subsequently closed down by the then minister for the environment, the leader of the Green Party in the French Parliament. Like so many political decisions, such actions may make sense in the short term, but in the long term may prove disastrous. When fast reactors are needed, they will not be able to compete with countries like Japan that are still developing them. India has no option but to accelerate its development of fast reactors to use the thorium that is plentiful there. It has had 13 MW fast reactors operating since 1985, and plans to build a prototype 500 MW fast reactor. China is building an experimental fast reactor of 25 MWe, to be followed by one of 600 MWe (Nuclear Issues 27, August 2005). The French have now decided to build both a sodium-cooled and a gas-cooled fast reactor to be ready for commercial use by 2035–2040. Russia has a 600 MW fast reactor at Beloyarsk and is building another of 800 MW (Nuclear Issues 28, December 2007).

Existing reactors are continually being improved and several new types of reactors are being studied. In the present thermal reactors the spent fuel rods still contain about 96% of the original uranium together with 1% plutonium, as well as the unwanted fission products. The utilisation of uranium can be improved by using uranium-plutonium mixed oxide (MOX) fuels. In France 1150 tonnes of spent fuel is being reprocessed in this way and converted into stable oxide form as a strategic reserve. One tonne of plutonium in the form of MOX fuel has an energy equivalent of two million tonnes of coal. This fuel will be used in the next generation of boiling water and pressurised water reactors.

Reactors with uranium as fuel and heavy water as the moderator were first built in Canada, and developed to form the CANDU power reactors. They are very reliable and economical in fuel since the moderator captures very few neutrons. The neutrons which are captured by the uranium 238 convert it to the fissile plutonium 239, and the fuel burn-up is nearly compensated by the plutonium 239 production, thus increasing time of operation before the fission fragments have to be removed. It is also possible to use a mixture of uranium 235 and thorium 232, and then the neutrons captured by the thorium 232 convert it to uranium 233 which is a more efficient fissile material than either uranium 235 or plutonium 239. The uranium 233 emits more neutrons per fission and makes it possible to use over 90% of the fissile material in the reactor.

Among the new reactors is the AP1000 already approved by the US Nuclear Regulatory Commission. This is essentially an improved version of the very successful pressurised water reactors (PWR) already operating in many countries. The improvements include a high degree of passive operation, using natural effects such as air circulation and gravity feed of cooling water, and a simplified design with fewer valves, pumps and pipes, and ventilation and cooling units. It occupies far less space and produces only 10% of the waste compared with the older plants (Nuclear Issues 26, September 2004). These reactors cost $US 1400 per kW (Speakers' Corner, October 2001).

Plans for the future include systems with self-sustaining fuel cycles such as sodium metal-cooled fast reactors, very high temperature reactors, supercritical water-cooled reactors, lead alloy-cooled reactors, gas cooled fast reactors, and molten-salt reactors. Some of the more promising of these projects may become available around 2030 (Goddard 2006).

Plans are also being made for other types of reactors, such as the pebble-bed reactor developed in South Africa following a German design. This is a relatively cheap low-power reactor that is inherently very safe. A Chinese energy consortium plans to build a 195 MW gas-cooled reactor that could be operating by 2010. This would put China in the leading place in a technology that could offer a new meltdown-proof alternative to water-cooled nuclear power stations. The world's only test pebble bed reactor is operating in the Institute of Nuclear and New Energy Technology in Beijing and provides the new technology for the planned reactors. In the United States, Westinghouse is considering leading a consortium to build a $500m PBMR in Idaho.

Another possibility is the accelerator-driven reactor that consumes its own waste, as discussed in the previous section. Detailed studies of its feasibility and cost are being made in Switzerland and in Los Alamos in the USA.

The continuing operation of nuclear reactors depends on the supply of uranium. The present economically recoverable uranium amounts to about 3.1 million tonnes and at the current rate of use amounts to about 70,000 tonnes per year, so that present resources are enough to about 44 years. It is estimated that highly probable deposits amount to at least four times that amount. Furthermore there are large deposits of uranium that are at present not economically recoverable, but can be used if needed. Doubling the price increases the recoverable reserves by a factor of about four. Introducing the fast reactor increases the uranium reserve by a factor of about sixty. Since the cost of fuel is only about 10% of the total running costs of a reactor, an increase in the cost of uranium would not appreciably affect the cost of nuclear power. There are also large amounts of uranium in granite (4 ppm) and seawater (0.003 ppm) contains about 4.5. billion tons.

The amount of uranium in sea water is so large that it is virtually inexhaustible, but it is not practicable to extract it (Norman, Worrall and Hesketh 2007). A technique for extracting uranium from seawater has been developed in Japan, at a cost of $100 per kilogram compared with the current cost of $20 per kilogram. The use of uranium obtained in this way would not appreciably affect the cost of nuclear power, but there are formidable difficulties in making the large amounts needed for a reactor programme. The Japanese method required 350 kg of absorbent material and collected 1.6 kg of uranium per year whereas a 1 GW reactor requires 160,000 kg per year. In addition even larger amounts of other salts in seawater would be produced. It is therefore doubtful if it is practicable to scale up the process to give the quantity of uranium needed.

In addition to uranium, there are four times more reserves of thorium. Although it is not itself fissile it can be converted in a reactor to fissile uranium 233. The Indian nuclear power programme plans to use the thorium-rich monazite sands in Kerala. They hope to have the first reactor using uranium-plutonium fuel with a blanket of thorium to breed uranium 233 in operation by 2010 (Nuclear Issues 26, January 2006; Ibid. 29, November 2007). Since all the thorium can be burned the total available energy is comparable to that obtainable from fast reactors (Mackay 2008, p. 106).

There is thus no danger of the development of nuclear power being curtailed by shortage of uranium or thorium in the foreseeable future.

4.8. Fusion Reactors

The most stable nuclei are those of medium weight around iron in the periodic table. This implies that we can obtain energy either by breaking up very heavy nuclei, as in fission, or by combining light nuclei. The latter possibility is the basis of the fusion reactor. The sun is a giant fusion reactor that produces its heat by chains of nuclear reactions that have the net result of combining four hydrogen nuclei to form an alpha-particle of mass four. It is not practicable to produce energy in this way on earth, but it can be done by raising hydrogen to a very high temperature as in the hydrogen bomb. To be useful as an energy source, the energy of fusion must be released in a controllable way. This may be done using reactions between deuterons and tritium, hydrogen isotopes with mass two and three. The reactions are:

DD: deuteron + deuteron ------- helion + neutron + 3.3 MeV.
and   ------- triton + proton + 4.0 MeV.
DT: deuteron + triton   ------- ---- alpha-particle + neutron + 17 MeV.

The main difficulty in getting these reactions to take place is that the deuterons and tritons are electrically charged and so repel each other. The attractive nuclear forces only come into play when the nuclei are very close to each other, and this only happens when they collide at very high energies. Since it is impracticable to aim each one individually, the only way to achieve this is to heat the hydrogen isotopes to very high temperatures of over a hundred million degrees. This is done in the hydrogen bomb by using the fission reaction as a detonator.

The DT reaction is preferred for fusion reactors because it has a higher energy yield and a lower ignition temperature (about a hundred million degrees) compared with that of the DD reaction (about three hundred million degrees). The fuel required is thus a mixture of deuterium and tritium. Deuterium is present in ordinary water at a concentration of about 1 in 6000, so there is practically unlimited supply. Tritium is unstable, with a half-life of about twelve years, so it has to be made. This can be done by bombarding lithium with neutrons, and this could take place in an absorbing blanket around the reactor. In this way it could produce all the tritium it needs, just like a breeder reactor. The lower ignition temperature means that it is likely that the first successful fusion reactor will use a deuterium-tritium mixture. Later on, when new designs can reach higher temperatures, the DD reaction can be used. We will then have enough energy to last for the foreseeable future, because the fusion energy in a gallon of water is ten times that in a gallon of petrol.

In the hydrogen bomb the fusion energy is released explosively. To harness this energy in a controllable way requires a way of releasing the energy slowly. At the very high energies of the fusion reaction the hydrogen is in the form of a plasma, with all the electrons stripped away from the atoms. For a fusion reaction to occur the plasma must be kept together at a high temperature long enough for the fusion to take place. Since the hydrogen nuclei are charged, the plasma can be contained by a magnetic field. In such a field a charged particle follows a curved path and one might hope to design a magnetic field so that the hydrogen nuclei go round closed orbits and never escape. Many attempts have been made to design a 'magnetic bottle' to contain the plasma, but it was found that instabilities always develop, preventing the reaction from taking place. A promising way to overcome this difficulty is to use a metal container in the shape of a torus or doughnut, and to put it in a magnetic field generated by electric currents passing through coils wound around the torus. Such fusion devices have been built at Princeton in the USA, at Culham in Britain as well as in Russia and Japan.

To produce a fusion reaction in such a device, a high vacuum is made in the torus and a mixture of deuterium and tritium injected into it. The mixture is then heated electrically so that the hydrogen nuclei collide violently with each other, and are prevented by the magnetic field from colliding with the walls of the torus. The energy produced when a deuteron collides with a triton is mainly carried away by the neutron. These neutrons are unaffected by the magnetic field and enter the surrounding blanket of absorbing material. They give up their energy to the absorber, raising it to a high temperature, and this can be used to produce steam and hence to generate electricity. Some of this electricity can be used for the initial heating of the deuterium and tritium.

The existing fusion devices are not able to confine the plasma for long enough for there to be a net energy gain. At Culham laboratory near Oxford, the Joint European Torus (JET) has been able to confine the plasma for a few seconds and to generate 16 MW of electricity (although 25 MW were needed to do this), so the goal is in sight. It is hoped that this will be achieved by a new fusion device called ITER (International Thermonuclear Experimental Reactor) that is being designed by a collaboration between China, India, Europe, Japan, Korea, Russia and the USA. This should give 500 MW for an input of 50 MW with a 400 second pulse length at an estimated cost of $12B. The design studies are quite far advanced, but it proved difficult to agree on where the reactor should be built, the main contenders being France and Japan. Eventually Cadarache in France was chosen, with Japan allocated the materials testing centre and the first prototype commercial fusion plant (Physics Today, March 2006).

ITER is a toroidal magnetic confinement system containing a plasma occupying 800 cubic metres controlled by a magnetic field of 5.3 T. It should show the deuterium-tritium plasma in conditions where the heating from the emitted alpha-particle is greater than the applied heating. Extensive design studies are in progress to determine the optimum conditions for the attainment of the fusion reaction. It is hoped that construction will start in 2008 and, if all goes well, the reactor should be operational by 2020, and then a larger machine can be designed that is planned to operate around 2040, with a commercial fusion reactor around 2060.

It is often said that fusion reactors are clean compared with fission reactors. Certainly it produces no fission fragments, but much radioactivity is generated by the neutrons entering the surrounding blanket, which accordingly has to be designed to reduce the radioactivity as much as possible. There are other hazards associated with the operation of fusion reactors, such as those due to the very high energies stored in the magnetic field, which is around 10 (11) Joules. There is also the danger of a lithium fire and of the escape of radioactive tritium.

It is also possible in principle to initiate a thermonuclear reaction using electron or ion beams from several directions focussed onto DT pellets. Laser beams can also be used. The initial effect of the beams on the pellets is to vaporise some of the surface and to set up an imploding shock wave that helps the fusion reaction. Some experiments have shown that fusion can be achieved in this way, but so far the energy required to power the lasers vastly exceeds that produced by fusion.

When a design has been found that produces a net gain in energy, then it will be time to design a fusion reactor for full-scale energy generation. Experience with fission reactors shows that it takes about twenty years from the establishment of technical viability of a reactor system to its large-scale implementation for power production. Fusion power is so important in the long run that research should be pushed forward rapidly, but until fusion power stations are actually in operation it is unrealistic to include them in any discussions of the energy crisis.

References

Cohen, B.L. (1999) Validity of the linear no-threshold theory of radiation carcinogenesis at low doses, *Nuclear Energy* **38**, p. 157 (London: The British Energy Society).

Darby, S., Doll, Sir Richard *et al.* (2005) Radon in homes and risk of lung cancer: Collaborative analysis of individual data from 13 European case-control studies, *British Medical Journal* **330**, p. 223.

Doll, Sir Richard, Evans, Dr. H.J. and Darby, Dr. S.C. (1994) Paternal exposure not to blame, *Nature* **367**, p. 678.

Goddard, Tony (2006) A future for nuclear power, *Physics World* (April), p. 15.

Hodgson, P.E., Gadioli, E., and Gadioli Erba, E. (1997) *Introductory Nuclear Physics* (Oxford: Clarendon Press).

Hodgson, P.E. (1999) *Nuclear Power, Energy and the Environment* (London: Imperial College Press).

Hodgson, P.E. (2002) *The Roots of Science and its Fruits* (London: The St. Austin Press).

Marshall, Walter (ed), *Nuclear Power Technology* (3 vols) (Oxford: Clarendon Press).

Murray, Raymond L. (1993) *Nuclear Energy* (4th ed.) (Oxford: Pergamon Press).

Norman, P., Worrall, A., and Hesketh, K. (2007) A new dawn for nuclear power, *Physics World* (July), p. 25.

Ott, Karl O., and Spinrad, Bernard I. (1985) *Nuclear energy: A Sensible Alternative* (New York and London: Plenum Press).

Taverne, Lord (2004) Speech in the House of Lords. SONE No.71.

Thompson, Richard E., Nelson, Donald F., Popkin, Joel H., and Popkin, Zenaida (2008) Case control study of lung cancer risk from residential radon exposure in Worcester County, Massachusetts Health Physics Society.

Van der Zwaan, B.C.C. (ed) (1999), *Nuclear Energy: Promise or Peril* (Singapore: World Scientific).

Weinberg, Alvin M. (1995) *The first Nuclear Era. The Life and Times of a Technological Fixer* (New York: American Institute of Physics).

Chapter 5

The Safety of Energy Sources

5.1. Introduction

All methods of generating energy have wider effects on the human community. They are all hazardous to some degree in different ways, and these hazards have to be taken into account when choosing which energy sources to use. The same applies to their effects on the climate and the pollution of the environment. These three subjects are considered in this chapter and in the two following chapters.

5.2. Comparative Safety

There is no completely safe way to produce energy. Coal mining is notoriously dangerous, oil wells catch fire, tankers collide or explode, dams burst and nuclear reactors burn and emit radioactivity. The renewable energy sources are sometimes described as safe or benign, but if we take into account the risks involved in manufacturing the equipment in factories and in constructing and maintaining them, it turns out that they are not so safe after all.

To make an objective assessment of safety it is necessary to include all the risks for each energy source. Mining, transport, construction, operation, maintenance and distribution all involve risks. Some are direct and affect only the workers, while some like pollution affect the whole population. The results of two studies are shown in Figure 5.1 and Table 5.1. The latter gives the number of deaths associated with the production of 1000 megawatt years of electricity by different energy sources. The numbers of injuries are larger and of similar relative proportions. There has been much controversy over these estimates, and they are subject to considerable uncertainties. It is not possible to produce better estimates because of widely varying conditions in various

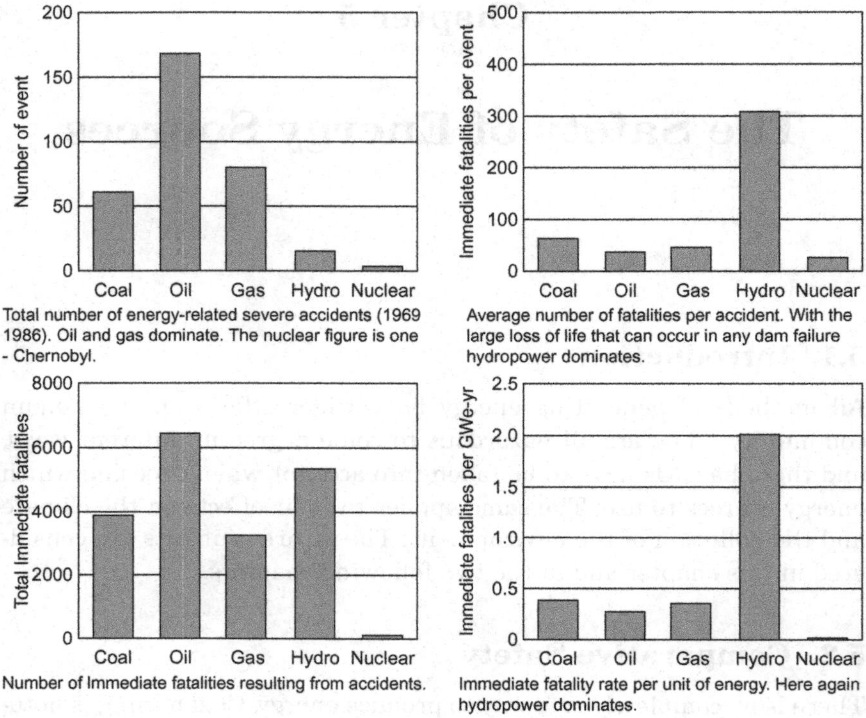

Total number of energy-related severe accidents (1969 1986). Oil and gas dominate. The nuclear figure is one - Chernobyl.

Average number of fatalities per accident. With the large loss of life that can occur in any dam failure hydropower dominates.

Number of Immediate fatalities resulting from accidents.

Immediate fatality rate per unit of energy. Here again hydropower dominates.

Figure 5.1. Accident statistics for various energy sources 1969–1986 (Paul Scherrer Institute 1995; *Nuclear Issues*, April, 1982).

countries and the gradual increase in safety with time. However even if the numbers are approximate, it is likely that the ranking order of the figures is essentially correct.

At first sight some of these figures may seem surprising. Coal is very hazardous because of the dangers of mining and also because of the large amounts of greenhouse and poisonous gases released into the atmosphere. It is estimated that there are about 16,000 deaths each year due to the emissions from coal power stations in the United Kingdom alone (Nuclear Issues 22, September 2000). These would not have occurred if they had been replaced by nuclear power stations. From 1976 to 1982 there were 1102 deaths due to oil tanker accidents that often also led to massive oil spills and extensive pollution. In addition there have been many fires on oil rigs. Gas leaks cause fires and

Table 5.1. The Average Numbers of Deaths and Days Lost Through Injuries Associated with the Production of 1000 MWy of Electricity (Inhaber 1981)

Energy Source	Deaths	Man–Days Lost
Coal	40	1000
Oil	10	400
Nuclear	1	8
Gas	0.3	3
Hydroelectricity	3	40
Wind	5	70
Wood	20	160
Solar (space)	8	70
Solar (thermal)	5	100
Solar (photovoltaic)	5	70
Ocean Thermal	2	25

explosions that were responsible for 39 deaths in Britain in 1989 alone. This may be compared with 366 in coal mining and 79 on off-shore oil and gas installations and five in the nuclear industry (not due to radiation). Wind and solar are more dangerous than we might expect because very many collectors have to be built, implying mining and construction hazards.

The accidents to oil tankers are particularly serious not only because of loss of life but for the serious damage they cause to the environment. The first of these disasters was when the Torrey Canyon ran on to rocks on 18 March 1967. This released 210,000 barrels of crude oil and this spread over 300 km of beaches in southern England and northern France. It was destroyed by the Royal Air Force to prevent loss of the remaining oil. On 16 March 1978 the Amoco Cadiz ran aground and spilled 1.5 million barrels of crude oil, causing extensive pollution. Other oil tankers such as the Erika on 12 December 1999 broke up in stormy weather. Oil is also spilt when drilling platforms are destroyed; 850,000 barrels of oil came from the Nowruz field in Iran on 4 February 1983. The blow-out of an exploratory well in the Gulf of Mexico spilled 3.3 million barrels. The destruction of wildlife is often severe. When the Exxon Valdez spilled 250,000 barrels in Alaska it covered 12,400 square kilometres with oil and killed over 34,000 sea birds, 10,000 sea otters and over sixteen whales. The cleanup operation cost $2.5 billion

(Aramco 2003). Fortunately, the environment recovers remarkably rapidly from oil spills, and the cleanup operations sometimes do more harm than good. Between 1992 and 1999, 77 tankers were lost at sea. There is also the hazard of leaks from oil pipelines. In 1989 a gas explosion in Russia was ignited by an electric train and about 575 people were killed and 600 injured (Nuclear Issues 29, June 2007).

It is notorious that it is the spectacular accidents that make the headlines and capture the public imagination, although over a long period of time (apart from dam bursts) they contribute relatively little to the total hazard of a power source. This effect is very familiar in other contexts; a train crash causing some tens of deaths receives more attention than the thousands killed individually in road accidents.

Such spectacular accidents are also a tragic feature of energy generation, and some statistics of the most notable in the period 1969–1986 are listed in Table 5.2. Mining accidents are particularly serious, and over 80,000 miners were killed in accidents from 1873 to 1938, and many hundreds of thousands more had their health permanently impaired by silicosis and other diseases. About 6000 people were killed in the Chinese coal industry in 2005 alone. Dams are less stable than they appear; water may trickle through cracks in the dam until finally the whole structure is undermined and it collapses. Such dam bursts can cause thousands of deaths among people living in villages below the dam. Another hazard is that a mountainside near the dam may be so weakened by heavy rain that it slides into the water held back by the dam; the displaced water then pours over the dam and may even destroy it.

Table 5.2. Severe Accidents in the Energy Sector 1969–1996 (Hirschberg et al. 2001). (Nuclear Issues 23, June–July 2001).

Energy Option	No. of Events	Max Fatalities	Max Fatalities Per Event	Max Fatalities Per GWe/Year
Coal	187	8272	44	0.342
Oil	334	15623	47	0.418
Natural gas	86	1482	17	0.085
LPG	77	3175	41	3.729
Hydro power	9	5140	571	0.884
Nuclear	1	56	56	0.008

The hazards of energy production can be reduced by more safety precautions. Thus many of the poisonous gases can be removed from the effluent gases of coal power stations by installing special equipment, as described in Section 2.2. Ideally, the cost of this should be set against the cost of the health effects of the gases if they are released. The safety devices themselves have to be manufactured, and this introduces further hazards. There comes a point when more safety devices do not have the net effect of reducing overall safety. A sensible compromise can only be reached by careful evaluation of all the hazards.

Wind power is relatively dangerous because large numbers of windmills have to be made to equal the output of one 1000 MW power station, and all industrial processes are dangerous in one way or another. In addition there are the hazards of erecting and maintaining the towers and the turbines. The huge rotors are liable to be damaged by gales and lightning strikes. In 2002 German insurance companies faced a bill of 40 million euros for such damages. There is little experience for offshore wind farms, but the insurance for them will certainly be greater than for those on land (Nuclear Issues 25, May 2003).

Extremely small amounts of radioactivity can be detected, and so it is easy to find any unwanted discharges and apply corrective measures. Strict limits to radiation exposure are enforced in nuclear installations as described in the previous chapter. Due to public pressure, large sums are spent on reducing minuscule hazards still further, although may more lives would be saved by spending the money in other ways.

As a result of the accidents to nuclear reactors at Three Mile Island and at Chernobyl there is widespread fear that all nuclear reactors are inherently unsafe and are liable to explode at any moment, spreading deadly radioactivity around the country. That at Three Mile Island was relatively minor and if the reactor had been left alone it would have corrected itself. Very little radioactivity was released and there is no evidence that it caused any harm, although the psychological and financial effects of the accident were severe. The accident at Chernobyl was a major disaster due partly to poor design and construction, and mainly due to gross disregard of the operating instructions by the controllers. Such a design would never be accepted in the West and now reactors have an excellent safety record. It is often said that the reactor at Chernobyl exploded. This is misleading if it is taken to mean that the reactor exploded like an atomic bomb. What happened was that when the safety circuits were switched off (contrary to rules) the reactor went prompt critical, and there was a surge of power that produced intense heat that blew off the cover of the reactor and set fire to the graphite moderator. This fire burned for several days and sent a large amount of radioactivity into the atmosphere. Some of the workers in the vicinity of the reactor received high doses of radiation, and of these 56 died. The much larger figures often quoted have no objective basis. There was no evidence of excess cases of leukaemia or other types of cancer among the thousands of workers employed in cleaning-up operations after the accident. The only effect was increased rates of thyroid cancer among children due to drinking water contaminated with iodine 131; this form of cancer can be treated effectively (Nuclear Issues 28, April 2006).

It has been estimated that the average risks to the population of the United States from a hundred nuclear power plants due to nuclear reactor accidents and radiation emission is less than four deaths per year, making the very questionable assumption of a linear dose relation between cancers and dose for very small doses of radiation. If there is a threshold dose, this number is reduced to a negligible level. This figure may be compared with the 300,000 deaths per year due to cancers from other causes, and 50,00 from automobile accidents.

It is also instructive to compare the casualties attributable to energy production with those due to natural disasters. Thus, for example, the Chinese Seismological Bureau estimated that in the years from 1949 to 1976, about 27 million people died and 76 million were injured following about a hundred earthquakes (Officer and Page 1993).

References

Aramco Magazine (2003) Winter, p. 35.

Hirschberg, S., Spickman, G., and Dones, R. (2001) *Severe Accidents in the Energy Sector* (Paul Scherrer Institute, Villigen, Switzerland).

Inhaber, Herbert (1981) *Risk of Energy Production* (Ottawa: Atomic Energy Control Board).

Officer, Charles, and Page, Jake (1993) *Tales of the Earth: Paroxysms and Perturbations of the Blue Planet*. [Quoted by John Carey in *The Faber Book of Science* (1995)] (London: Faber and Faber).

Chapter 6

Pollution of the Environment

6.1. Introduction

We are increasingly conscious of the effects of our activities on the natural environment. Man has always affected the environment to some extent, but this has become a serious matter of public concern during the last century. These effects may be classified according to the area used or affected by the energy generators, and by the pollution they produce. This pollution may be local or global, and includes not only poisonous chemicals, but also visual and aural pollution, and affects the atmosphere, the land and the sea.

When considering man-made pollution it is useful in appropriate cases to put it in perspective by comparing it with the natural sources of pollution that are beyond our control. Thus bush fires due to lightning strikes have always occurred, and are even necessary for the germination of some plants. Volcanic eruptions throw huge amounts of poisonous chemicals into the atmosphere, and this falls on land and sea. The earth has great natural recuperative powers and, once the source of the pollution is removed, the land, lakes and seas return to their previous state.

6.2. Pollution of the Land

The area affected by energy generation depends on the degree of concentration of the source. The energy in the diffuse sources such as wind and solar is widely spread and so collectors occupy a large area. Some sources, such as wind, require the generators to be sited on high ground, and so the area they affect is much larger than that occupied by the collectors themselves, and so is the corresponding visual and aural pollution.

Most people dislike the intrusive presence of many wind turbines, especially in areas of great natural beauty. The generators also emit a persistent humming noise which many people living nearby find intolerable, and results in the loss of value of their property. These disadvantages do not occur in the case of off-shore wind farms, but instead there is some danger to shipping and appreciably higher cost. Similar remarks apply to solar power, if ever the collectors were deployed on a large scale.

It is sometimes argued against wind power that the turbine blades kill large numbers of birds, estimated to be about 30,000 per year in Denmark and 70,000 per year in the USA. While this is of course regrettable, it may be put into perspective by comparing these numbers with the numbers of birds killed on motorways (a million a year in Denmark and 57 million per year in the USA), by colliding with glass windows (98 million per year in the USA), by domestic cats (55 million per year in Britain) (Lomborg 2004) and by massive habitat loss such as that at Saemangeum in Korea (Brown 2004).

In the case of hydropower, the collector is a mountain valley that remains unaltered, but the lake where the rainfall is collected inundates a large area, often of agricultural land that may also include ancient villages and valuable wilderness areas. In times of low rainfall ugly bands of mud may be exposed around the lake. The concentrated energy sources that generate energy inside themselves occupy far less land and, although they are large structures, they can be unobtrusively sited. Some comparative figures for the land occupied by various energy sources are shown in Table 6.1.

Table 6.1. Areas of Land Occupied by Various Power Sources in Square Metres Per Megawatt

Wind	1,700,000
Hydro	265,000
Solar	100,000
Coal	2400
Gas	1500
Oil	870
Nuclear	630

6.3. Pollution of the Atmosphere

Energy generation also pollutes the atmosphere as well as the land and the sea. Several estimates of the atmospheric carbon dioxide emissions are given in Table 6.2. They agree reasonably well and show very clearly the overwhelming preponderance of emissions from the fossil fuels coal, oil and gas. In sharp contrast, the emissions from nuclear, wind and hydro are only around 1% of those due to the fossil fuels. The relative emissions due to various sources in the European Union are: oil 50%, coal 28% and gas 20%. In terms of consumer sectors electricity generation accounts for 37%, transport 28%, industry 16%, household 14% and the service sector 5%. The emissions from transport are expected to increase by 50% between 1990 and 2010 (Nuclear Issues 23, March 2001). Worldwide, the relative emissions are oil 42%, coal 37% and natural gas 20%. Between 1990 and 2001 world carbon dioxide emissions increased from 21,563 to 23,899 million tonnes (US Department of Energy Outlook for 2004; Nuclear Issues 27, March 2005).

There are notable differences in the emissions in various countries, as shown in Table 6.3.

The emissions of carbon dioxide from various form of transport in grams of carbon dioxide per 100 passenger-km are 278 for an average 1990 car; 161 for a diesel car; 69 for a bus; 79 for a diesel train; 76 for

Table 6.2. Atmospheric Carbon Dioxide Emissions in g/kWh

Energy Source	OECD	BNFL2001	Vattenfall*	NI2007
Coal	955	955	980	750–800
Oil	828	818	—	550
Gas	430	446	450	400–440
Nuclear	4	4	3.1	15
Wind	8	7	5.5	10–30
Hydro	8	4	—	5–20
Biomass	17	—	—	30
Geothermal	79	—	—	—
Solar	133	—	—	—

*Vattenfall is the Swedish State Power Board (Nuclear Issues 28, January 2006).

Table 6.3. Atmospheric Carbon Dioxide Emissions in g/kWh (Kivisto 2000)

	Japan	Sweden	Finland
Coal	975	980	894
Gas (Thermal)	608	1150	—
Gas (combined cycle)	519	450	472
Solar photovoltaic	53	50	95
Wind	29	5.5	14
Nuclear	22	6	10–26
Hydroelectric	11	3	—

Table 6.4. Values of the Kyoto Index for the UK in 1999 (Nuclear Issues 22, November 2000)

Source	Electricity Generated	GWh Carbon	Mte Kyoto Index
Coal	106072	26.52	0.5
Oil	5551	1.11	0.6
Gas	141365	17.67	1.0
Nuclear	96281	0.24	50
Biomass	8377	0.10	10
Hydro	5352	0.03	21
Wind	898	0.005	24

an electric train and 853 for aircraft, assuming 80% occupancy for aircraft and 40% for the other modes of transport (MacKay 2008, p. 135).

The relative purity of the emissions from different energy sources can be measured by the Kyoto Index, which is the percentage share of generation divided by the percentage share of carbon emissions. Values of this Index for emissions from the UK in 1999 are given in Table 6.4.

The pollution due to the fossil fuels is partly the carbon dioxide that is a natural product of the burning fuel and partly a wide range of poisonous chemicals due to impurities in the fuel. This latter pollution depends on the constituents of the fuel, and some typical figures for coal are given in Section 2.2. Much of this returns to the earth in the form of acid rain, and is responsible for the destruction of forests and the poisoning of seas and lakes. In the 1920s and 30s lakes in Scandinavia lost most of their salmon and trout and by the 1980s, 4000 lakes were almost dead and 5000 had lost most of their fish. Other forms of aquatic life and birds were also affected, and conifers, beech, oaks and spruce in the forests of Europe and North America were observed to be dying (Christianson 1999).

Initially it was suspected that the effects on the trees were due to some virus, but measurements of the acidity of the rainwater showed that it was due to industrial effluent from factories hundreds of miles away. It was thought that pollution could be avoided by building taller and taller factory chimneys, but this serves only to spread the poisons even more widely. The only solution to the problem of acid rain is to remove the pollution at its source.

Detailed surveys are necessary to assess the extent of the effects of acid rain and some studies have shown that the effects are comparatively small; for example, it has been estimated that acid rain affected only about 0.5% of the European forest area. Many lakes are still affected, though most are recovering due to reductions in the emissions of sulphur dioxide. In some cases, the sulphur and the nitrogen in the acid rain actually helped plant growth (Lomborg, op. cit. pp. 178–181). However, the pollution of the atmosphere due to sulphur dioxide, nitrous oxide and ozone from fossil fuels is estimated to be responsible for 330,000 deaths in 14 years in the UK.

Serious local pollution is caused by factories and energy sources pouring their effluent into the atmosphere or into lakes, rivers and seas. The effect of acid rain on lakes and land is determined by the buffering capacity of the geology of the region but in many cases it is severe. This hazard is now recognised, and stringent regulations have stopped many of these discharges, and then as a result the air has become cleaner, and fishes have returned to rivers and lakes. Thus in the nineteenth and early twentieth century millions of domestic coal fires produced the London smogs. The notorious example in 1952 killed about three thousand people, mostly among those already suffering from respiratory diseases. I recall that smog very well; it was easy to get lost in familiar streets, as this smog was so thick that one could not see one's feet. When one was directly underneath a street light it was just possible to see it through the gloom, and when walking along the road one had to walk in the general direction until one saw the next street light. Of course all traffic was stopped. Now that domestic coal fires are forbidden the air has become cleaner and such smogs can never occur again.

A paper issued by the European Community on energy strategy points out that nuclear power 'will make it possible for Europe to save about 330 million tonnes of carbon dioxide emissions in 2010'. 'This is equivalent to taking a hundred million cars off the road' (Nuclear Issues 23, March 2001).

The renewable energy sources emit practically no poisonous gases, except from those associated with the manufacture of the devices. The amounts of radioactivity emitted by several types of power stations are given in Table 6.5. These amounts are all very small compared with the natural background radiation, but it is nevertheless notable that coal power stations emit more radioactivity than nuclear.

Although localised pollution is usually dispersed until it is no longer noticeable, this simply means that it is spread throughout the atmosphere. Thus, as in the case of carbon dioxide, the average global pollution steadily rises. Apart from the special case of carbon dioxide, the level of the global pollution by heavy metals and other chemicals with the possible exception of some pesticides is still so low that it cannot cause appreciable harm on a global scale.

There are also some chemicals that can destroy the ozone layer in the stratosphere about thirty miles above the earth that absorbs the ultraviolet light that is harmful to some plants and insects and can cause skin cancer in unwary sunbathers. From the beginning of the twentieth century these chemicals were widely used as cleaning solvents, pest control and for fire protection. The major threat to the ozone layer came from the development of the chlorofluorocarbons (CFCs) in the nineteen thirties, and their widespread use in refrigerators

Table 6.5. Emissions of Nuclear Radiation from Various Power Sources in Man-Sieverts Per Gigawatt-Year (IAEA world average)

Coal	4.0
Nuclear	2.5
Geothermal	2.0
Peat	2.0
Oil	0.5
Gas	0.03

and aerosol propellants. Two chemists, Mario J. Molina and F. Sherwood Rowland, suggested in 1974 that the CFCs could undergo photodissociation in the stratosphere, producing chlorine that destroys the ozone layer. The chlorine acts as a catalyst in the reaction dissociating the ozone, so that a single chlorine atom can destroy many thousands of ozone molecules. In addition, in 1975, the chemist Veerabhadran Ramanathan showed that the CFCs are also efficient greenhouse gases. Not surprisingly, the CFC industry responded by denying these effects.

Confirmation eventually came through the work of Joe Farman of the British Antarctic Survey, who year after year made measurements of the ozone level with his spectrophotometer. By 1984 he had records stretching back for twenty-seven years, and he found that the ozone level progressively decreased in the years from 1982 to 1984 until they were 40% below the average value. He had found a hole in the ozone layer. He also had a theory about why it occurred in the Antarctic region, for there the seasonal increase in the intensity of the ultraviolet radiation renders the stratosphere over the Antarctic especially sensitive to chlorine. Farman's results were confirmed by analysis of data obtained by NASA's satellite Nimbus.

As a result of this work, the United Nations Environment Programme initiated intensive research on the subject. In addition, many companies looked for alternatives and started to reduce the emission of CFC's and other ozone-destroying chemicals, culminating in the signing of the regulatory Montreal Protocol in 1987. This encouraged manufacturers to develop alternative environmentally-friendly products and eventually to the significant reduction of ozone-destroying chemicals. The concentration of such chemicals in the atmosphere peaked in 1994 and is now declining and should eventually lead to the closing of the ozone hole (Anderson and Madhava Sarma 2002; Christianson 1999; Lomborg 2004, pp. 273–276).

Once again it is useful to put the results of the depletion of the ozone layer in perspective by comparing estimates of the number of skin cancers with and without the ozone hole. The numbers of cases of skin cancers increased greatly during the twentieth century, but since there is a long period of latency most of the increase must be attributed to exposures some decades before the depletion of the ozone layer. These exposures to harmful ultraviolet radiation increased over previous levels due to more sunbathing, increased life span and other causes.

The depletion of the ozone layer has certainly increased the exposure, and with it the number of cancer cases. Most skin cancers are easily curable and only about 5% are the more lethal melanomas. It is estimated that of about a million cancer cases each year those attributable to the depletion of the ozone layer will peak at about 27,000 cases per year in the year 2060 in the USA. In other words, the depletion of the ozone layer is responsible for an increase in the number of cases of about 3%. Since the intensity of the ultraviolet radiation depends strongly on latitude, the increased probability of contracting skin cancer due to the depletion of the ozone layer is the same as that incurred by moving about 200 km closer to the equator.

There is increasing public recognition of the importance of reducing carbon emissions into the atmosphere. To do this industrial companies are being encouraged to assess the carbon emissions from the whole production process from obtaining the raw material and manufacturing the product, to its distribution and eventual disposal. It is planned that the product will bear a label giving its carbon emission, so that the purchaser can take this into account. If all manufacturers do this, commercial pressures will stimulate them to reduce their carbon emissions. Such an examination of the whole production process could also produce significant fuel savings (Cave 2008).

It is useful to consider the minimum reduction of carbon dioxide emissions necessary to avert a serous crisis. MacKay (2008) notes that there is some agreement that global warming must be kept below 2°C. Any higher rise could have many highly undesirable consequences. For example, the Greenland ice cap would start to melt, thus raising the sea level by about seven metres over a long period. This would be sufficient to inundate many large cities, from London to Tokyo and from New York to Calcutta. To keep the temperature below this level requires the level of carbon dioxide in the atmosphere to be kept below about 400 ppm. This in turn requires each person to limit their emissions to about one tonne of carbon dioxide each year. Since the current emissions in the United Kingdom are in the region of ten tonnes per person per year, it is evident that drastic changes to our lifestyle are necessary.

In recent years there have been several conferences on the need to tackle global warming, and most countries have agreed to reduce their emissions by 10% to 15% over the next decade. There are strong disagreements about the actual amounts and how the burden should be shared between the developed and developing nations. It is becoming increasingly clear that these arguments are increasingly unreal, as the amounts under discussion only scratch the surface of the problem. To avoid disaster, all countries must recognise that they must strive to reduce their emission to zero as soon as possible. This will require vast and costly changes in our living styles, but the results of just making small changes will be far worse.

One way to reduce carbon dioxide emissions is by means of a tax on all processes leading to such emissions, and this is already in operation in several countries. This is a useful source of revenue, but to be effective the level of taxation needs to be high enough to alter people's behaviour. At present the tax is in the region of £7 to £21 per tonne of carbon dioxide. However, assuming that doubling the price of fuel would significantly affect car usage, the tax would need to be about £500 per tonne. To discourage air travel, a figure of around £400 per tonne is needed, and other activities would require similarly high taxes. Thus much higher taxes are needed to be effective. Low taxes that do not affect people's behaviour are worse than useless, as they give the illusion of helping to solve the problem and may even be counterproductive because they encourage the idea that paying the taxes removes the damage (MacKay 2008).

6.4. Pollution of the Sea

The transport of fuel from source to power stations is often hazardous. Coal is easily transported by train or by ship, and coal power stations are often sited near the coalfields. Oil is easier to transport, but the hazards are greater. Over the years, many huge oil tankers have run onto rocks and large amounts of oil have escaped, with devastating consequences for the local environment. Large numbers of fish and seabirds are killed, and marine plant life destroyed. However, the imposition of stringent safety regulations has reduced the number of such accidents, and the effects are generally rather short-lived. Gradually, the oil evaporates, is broken down naturally, or finally becomes harmless pieces of tar. Detailed studies have shown that the polluted area often returns in many respects to its previous state more quickly than many thought possible, and that expensive attempts to clean the beaches often delay their eventual regeneration. It should also be mentioned that only about 20% of the oil entering the oceans comes from such oil spills.

Most of the remainder comes from the common practice of filling empty tanks with water as ballast and then pumping it out again when it is ready to receive oil. Finally, some comes from natural oil seepage from oil-bearing rocks.

6.5. Radioactive Pollution

The radioactive pollution of the environment comes partly from nuclear and coal power stations and partly from other industrial processes. That from nuclear power stations has been discussed in Section 4.5 and those from various power sources are given in Table 6.2. In addition many industrial activities use naturally occurring minerals that contain varying amounts of radioactive material. Such industries include those producing phosphorus and phosphoric acid, fertilizers, iron and steel, coke, minerals, coal tar processing, ceramics and uranium and thorium mining (Nuclear Issues 22, November 2000). Some of this radioactivity remains in the product and the remainder is discharged into the sea or buried underground. The total amount of radioactivity discharged into the oceans is estimated to be about 0.1 EBq, compared with the amount already there of about 10,000 EBq.

Most of these activities release very small amounts of radioactivity and in any case there is little that can realistically done about them. In this context very small means far less than the natural background to which we are all exposed, which can vary by factors as high as a hundred from one place to another without there being any detectable effects in the places with the higher amounts of radioactivity. There is a tendency to impose strict limits on the emissions from the nuclear industry while accepting far higher amounts from unavoidable sources. As a result, the recommended exposure limits for various situations vary from 0.01 to over 100 ms/y.

For the Nordic countries the collective dose from emissions from nuclear reactors is about 20 person-Sv/yr for workers and one person-Sievert/yr for the general public. The emissions from coal in Denmark and peat in Finland give doses of about 80 person-Sievert/yr. In addition, the coal ash, amounting to 280 million tonnes per year, if used as building material, bricks and cement, road stabiliser, asphalt mix and fertilisers, giving doses up to several mSv (Nuclear Issues 23, June/July 2001).

6.6. The Greens

In many countries Green Parties have been established and work energetically to conserve the environment. They have fought often bitter battles against logging companies, dam builders and Governments to preserve the natural environment. Sometimes they have been successful as in their campaign to save the Franklin River in Tasmania; sometimes they have failed as in the case of Lake Pedder. Frequently they have been abused, shot at and attacked by their opponents, and on some occasions the forces of law and order have failed to protect them. No one who reads the book 'Memo for a Saner World' (Brown 2004) could fail to admire their efforts and be horrified at the opposition they have encountered as they fought to preserve the environment in many countries. This book also contains a copy of the Greens Manifesto, and included among many admirable statements that strongly support the renewable energies and condemn nuclear power. As explained in the previous chapter, however desirable the renewables may be, they are simply quite unable to provide the energy we need. So, as nuclear is non-polluting, reasonably safe and economic, it is very regrettable that the Greens are so implacably opposed to it. It is not ever discussed but instead is dismissed as unworthy of consideration. Of all the energy sources, nuclear is the one that least harms the environment, so it is particularly sad that it is regarded as an enemy and not as a friend.

References

Anderson, Stephen O., and Madhava Sarma, K. (2002) *Protecting the Ozone Layer* (The United Nations History. Earthscan).

Brown, Bob (2004) *Memo for a Saner World* (London: Penguin Books).

Cave, Andrew (2008) Taking good measures, *Daily Telegraph Business Supplement* **21** (March), p. B2.

Christianson, Gale E. (1999) *Greenhouse: The 200-Year Story of Global Warming* (New York: Walker and Company).

Lomborg, Bjorn (2004) *The Skeptical Environmentalist* (Cambridge: Cambridge University Press).

MacKay, David J.C. (2008) *Sustainable Energy: Without the Hot Air*.

Chapter 7

Climate Change

7.1. Introduction

By climate we mean the time averages of the many variables describing
the condition of the atmosphere: the temperature, purity and humidity of the air, the rainfall, the strength of the winds and storms, and the
clouds, mists and fogs. All these are constantly changing, and we can
describe it by taking averages for a local region or for the whole world.

It is useful to distinguish between the climate and the weather. The
weather is what we experience from day to day; it is always changing
and impossible to predict more than a few days in advance. This is
because the atmosphere is what is called a chaotic system; very small
changes can have large effects. Climate refers to averages of the
weather taken over periods of several years. These averages smooth out
the fluctuations of the weather, and are what we are concerned with
here. It is not always easy to detect long-term changes in a fluctuating
quantity, and this accounts for the difficulty of studying climate change.

During the last two million years studies of sediments, ice cores and
cave deposits have shown that there have been a succession of ice ages due
to changes of the earth's orbit. In the colder periods most of Europe was
covered by ice sheets up to 3 km thick. We are now in a relatively warm
period, when there are large temperature differences between the hot and
humid equator and the ice-bound poles. This temperature difference of
about 70°C accounts for much of our climate, as warm air from the equator moves towards the poles, stirring up hurricanes, typhoons and monsoons. Superposed on the larger changes have been shorter climatic
fluctuations that have strongly affected human history (Maslin 2004).

The changes in the climate over the last millennium have been
found by studying tree rings, ice cores and corals. The results are consistent, which confirms their accuracy. During the last forty years more

extensive data have been obtained by instruments carried aloft by balloons and by satellites. The most important long-term effects are changes in the average temperature and in the sea level. The average temperature has remained much the same for the first 900 years of the millennium, and since then has risen on the average by about 0.005°C per year. In the same period the sea level has risen by between 4 and 14 cm, the uncertainty being due to the rising or sinking of the land.

Climate is one of the determining features of civilisation, so any change in the climate can have momentous consequences. The importance of climate for civilisation is discussed in Section 7.2.

It is becoming increasingly evident that our present energy policies may be having a disastrous effect on the world climate. Already these have been estimated to cause 160,000 deaths per year due to heat waves, flooding and crop damage (Physics World, July 2007, p. 25). This is attributed to the carbon dioxide and other gases that are released when fossil fuels are burned. The chemical reaction that releases heat is the combination of the carbon in the fuel with oxygen in the air to produce carbon dioxide, and this is released into the atmosphere. The only way to avoid this is by a variety of chemical processes referred to as carbon sequestration described in Section 2.1. This process is so expensive that for the foreseeable future all the carbon dioxide from burning fossil fuels will be emitted into the atmosphere. As described in Section 7.3, there is increasing evidence that this is seriously affecting climate throughout the world.

Fossil fuel power stations also release many poisonous substances into the atmosphere that eventually fall to earth as acid rain, killing trees and changing the ecology of rivers and lakes. An IAEA study estimates the emissions from a coal power station to be (in kg/MWh) 830 for carbon dioxide, 2.16 for nitrogen oxides, 0.6 for sulphur dioxide and 0.1 for particulates containing radioactivity (Nuclear Issues 26, No. 8).

The effects of these emissions into the atmosphere are responsible for global warming, as described in Section 7.4. The effects of global warming are described in Section 7.5 and the possibility of catastrophic changes in Section 7.6.

7.2. Climate as a Determining Feature of Civilization

It is notable that over the last six thousand years many great civilizations have arisen, enjoyed a few centuries of power, and then faded into insignificance. Why has this happened when and where it did?

There are many conditions necessary for the rise of a civilization. The land area must be large enough to support at least a few hundred thousand people. The land must be fertile enough to support a range of edible crops, and there must be available animals that can be domesticated and are strong enough to carry burdens, including man himself (Diamond 1999).

In addition, the climate must be suited to man himself, not so hot and humid that he becomes lazy and listless, especially if there is plenty of fruit for the picking. On the other hand it must not be so cold that all his energies are concentrated on survival. Extensive studies have shown that man survives best at a temperature between 60 and 70 degrees F, with moderate humidity. He feels well, and is active and energetic. Below about 60 degrees it is too cold, but somewhat above 70 is quite tolerable. We might therefore expect the earliest civilizations to arise around the 70 degrees isotherm, and indeed this is the case. The isotherm passes through North Africa, Egypt, Mesopotamia and the Indus valley, all sites of early civilizations (Markham 1947).

Subsequently, improvements in technology enabled people to survive in cooler regions. They learned to make clothing to keep them warm and to build houses to retain the heat. Glass was invented and made windows possible. The Romans invented the hypercaust that enabled them to keep their homes warm in cold weather. Civilization then spread northwards with the help of Roman technology. Around the end of the first millennium AD improvements in agriculture further aided the spread of civilization. The stirrup and the horse collar enabled the horse to replace oxen for ploughing, and together with the wheeled plough that turned the soil over enabled the heavy fertile soils of Northern Europe to be opened up for cultivation. This in turn enabled the land to support a much greater population than ever before.

Further support for the influence of climate comes from the experiences of people who emigrate from a temperate climate to one that is hot and humid. The first generation survives well, but subsequent generations become less active and listless, leading eventually to the 'poor white' people living permanently in countries like the Bahamas and South Africa. Such people can sink lower than the much more primitive indigenous populations. This decline can be avoided if the emigrants take periods of leave to spend time in their home countries, as was the practice of many people who went to tropical climates such as India and the Far East.

Now, with heaters in the winter and air conditioning in the summer, we are able to survive and work in more extreme conditions and so the enervating effects of an unfavourable climate are not so pronounced, and may be eliminated entirely. Our global civilization has spread over most of the earth apart from the very cold regions around the poles and the deserts of Africa and Central Asia. We have largely become divorced from the natural processes on which ultimately we depend. We are not so conscious of the seasons, as we enjoy continuous supplies of fruits and vegetables from all over the world. If, however, some catastrophic climate change occurred, coupled with destructive storms and tidal waves, it might not be possible to rebuild the infrastructure that enables us to prosper in somewhat unfavourable climates.

We would do well to remember that our civilization depends on adequate supplies of energy and we are still at the mercy of earthquakes, tornadoes, tsunamis and volcanic eruptions. We still rely on nature to provide our food, and this in turn depends on the climate. If the climate changes, it may become more difficult to grow the crops we need, and our domestic animals may suffer. If the increasing demand for more energy is met by burning fossil fuels the resulting pollution will reduce the productivity of the land and the sea. At the same time the world population inexorably increases and the demand for a higher standard of living will be difficult if not impossible to meet.

These problems are inextricably linked together. The rising population and its rising expectations require more energy. The production of this energy is made increasingly difficult as the reserves of possible fuels are exhausted. Energy production can cause global warming and climate change, and this reduces the capacity of the earth to grow the needed food. In the end, something has to give way. Will it be famine, or wars or pestilence, or a combination of the three?

7.3. The Reality of Climate Change

It took a long time for the reality of climate change to be recognized, in particular the global rise in temperature at sea level. The story began with the pioneering work of the Swedish scientist Svente Arrhenius, who realized that industrial processes leading to the emission of carbon dioxide could warm the earth. This conclusion was supported by the independent research of Thomas Chamberlain. Their results attracted little attention because it seemed that human activities must be insignificant compared with many other effects on the climate such as the sunspot cycle and the changes of the earth's orbit around the sun. Furthermore, there is about fifty times as much carbon dioxide in the oceans than in the atmosphere and it was thought that they would easily absorb any extra pollution due to human activities (Maslin 2004).

During the nineteen fifties improved scientific studies of the atmosphere and the oceans showed that additional carbon dioxide in the atmosphere would warm the earth and that only about a third of this would be absorbed by the oceans. By about 1960 there was general agreement that the temperature of the earth should increase due to human activities. However this seemed to be contradicted by the overall fall in temperature from 1940 to 1975. Scientists then predicted continual cooling, leading to widespread famine due to floods and drought.

Subsequently, however, the global temperature began to increase and by 1990 the growth was unmistakable, although there were still grounds for doubt. In particular, there had been changes in the methods of measurement of temperature, the previous cooling was not understood and the satellite measurements of the upper atmosphere did not show the warming trend. Further work showed that for various reasons these indications are unreliable, and the steady increase of temperature was confirmed.

During the nineteen eighties people became more conscious of the importance of environment. This was due to many events, from the publication of Rachel Carson's book The Silent Spring in 1962, the Report of the Club of Rome on the limits to growth to the nuclear accidents at Three Mile Island and Chernobyl. The global nature of climate change was shown by the discovery in 1955 of the ozone hole in the atmosphere above Antarctica that was discussed in Section 6.3. In this case prompt international action in banning the use of chlorofluorocarbons succeeded in averting the danger.

In addition to these global changes there is impressive evidence for the reality of climate change during the last few decades. Some of this has been described in a book by Sir Ghillean Prance, former Director of the Royal Botanic Gardens in Kew in London. He recalls that there were devastating floods in Mozambique and Venezuela and quite serious ones in England. In other countries there has been drought, that in the Midwest of the United States in 1988–1989 caused losses estimated at \$39 billion. The hurricane Mitch killed ten thousand people in Central America. The average temperatures are rising in many countries: of the five warmest years ever recorded in the United Kingdom, four have been in the last decade. One result is that in some regions the growing season for plants is increasing, with earlier development in spring, and autumn events being delayed. Birds and animals are also affected, and some species, unable to cope with the climate change, have become extinct. Already British birds nest about twelve days earlier than they did a few decades ago.

For the last thirty years the isotherms have been moving steadily towards the North and South Poles at a rate of about thirty-five miles per decade. Some animals and plants are able to move and keep up with this, but others may find it difficult or impossible due to biological factors and man-made and natural barriers. Since they are in many ways interdependent this could disrupt the ecosystem, causing the extinction of some species. In the past, rises in temperature have caused mass extinctions of 50% to 90% of all species, and the same could happen again.

Such evidence raises many questions. Do these changes show that world climate is changing? If so, will it continue to change in the same way? Are these changes due to human actions? If so, what can we do about it?

Climate is determined by many natural causes, and in addition there is evidence that it is affected by human actions, as shown in Figure 7.1. We cannot do anything about the natural causes, but if there is a causal link between human actions and climate change we may have reason to expect the present changes to continue, and furthermore we will have a strong incentive to take action to mitigate the harmful effects of climate change.

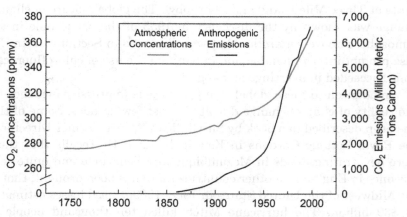

Source: Oak Ridge National Laboratory, Carbon Dioxide Information Analysis Center, http://cdiac.esd.oml.gov/.

Figure 7.1. The Effect of Anthropogenic Emissions on Atmospheric Concentrations of Carbon Dioxide.

A causal link between carbon dioxide emissions and global warming has been proposed; it is the greenhouse effect that is discussed in the next section.

7.4. The Greenhouse Effect and Global Warming

Extensive measurements have shown that the concentrations of carbon dioxide, methane and some other gases in the atmosphere are steadily increasing. In the 1780s the level of atmospheric carbon dioxide was about 280 parts per million, as it had been for the last six thousand years. Industrialisation increased the level to 315 ppm by the 1930s, 330 ppm by the mid-1970s and 360 ppm by the mid-1990s. In the last ten years the level has risen by a further 20 ppm. By the middle of the present century it could rise to 500 ppm. The annual increase of carbon dioxide is now 0.4%, that of methane 1.2%, of nitrous oxide 0.3%, of the chlorofluorocarbons 6% and of ozone about 0.25%. In the European Union, fossil fuels are the main source: oil 50%, natural gas 20% and coal 28%. Of this, electricity generation accounts for 37%, transport 28%, industry 16%, households 14% and the service sector for 5%. These are established facts, and in addition there is a strong correlation between carbon dioxide concentrations and temperature changes, as shown in Figure 7.2. It is then suggested that these increased concentrations are responsible for global warming and that global warming is responsible for other climate changes and predicted effects such as a worldwide rise in the sea level. The evidence for anthropogenic climate change has increased in recent years and its reality is now generally accepted.

The mechanism of global warming is called the greenhouse effect, though strictly speaking this is a misnomer. The sun's rays pass through the atmosphere and warm the earth. This energy is re-emitted at a longer wavelength that is scattered by the 'greenhouse' molecules in the atmosphere and some of this radiation is scattered back and warms the earth so that its average temperature is around 14°C, whereas without the greenhouse effect it would be around −18°C. If the concentration of the greenhouse gases increases, the earth becomes hotter. The 'greenhouse' molecules are primarily carbon dioxide, methane and the chlorofluorocarbons. In a greenhouse the glass transmits the incoming solar radiation and it warms the earth. The earth radiates energy but the radiation has a wavelength that cannot escape, and so the incoming heat is trapped, thus warming the greenhouse.

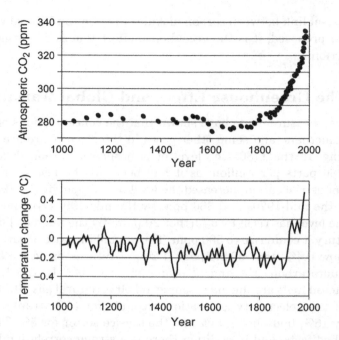

Figure 7.2. The Average Atmospheric Carbon Dioxide Concentration over the Last Millenium Compared with the Average Temperature Changes in the Northern Hemisphere (Nuclear Issues 23, May 2001).

The greenhouse gases differ greatly in their contribution to the greenhouse effect by their concentration, their scattering efficiency and the time they remain in the atmosphere. The concentration of carbon dioxide is relatively high and it remains for a long time and so it is the most important contributor to global warming. Methane has about sixty times the scattering power but has low concentration and a relatively short life of about twelve years, so it is not so important. This could change if the emission of methane increases due to global warming. The chlorofluorocarbons are extremely potent greenhouse gases and so their emission is now prohibited by international agreement. They are all due to anthropogenic causes and so all machines, such as refrigerators, that emit this gas have been re-designed to use other gases, thus eliminating this source of global warming.

The effect of methane on global warming will become more important due to melting of the permafrost in northern countries such as Alaska and

Siberia. Normally the ground is frozen to a depth of many metres, but an increase in the summer temperature causes the top layer to melt and emit methane as well as carbon dioxide. These gases enhance global warming, leading to further melting and emissions, in a positive feedback process. The melting of the permafrost will also cause mud slides and avalanches that could threaten houses, roads and pipelines (Maslin 2004).

An additional and potentially even more serious danger is the breakdown of gas hydrates. At very low temperatures and high pressures water molecules can form a solid consisting of cages holding molecules of methane and other gases. These are gas hydrates, and are found in deep oceans and beneath the permafrost. It is estimated that there are more than 10,000 gigatonnes of gas hydrates whereas there are just 180 gigatonnes of carbon dioxide in the atmosphere. If only some of this were released due to global warming it would add to the methane and carbon dioxide in the atmosphere. This would intensify the global warming, causing further releases and so on in another positive feedback effect. There is some evidence that such a runaway greenhouse effect took place 55 million years ago. About 1200 tonnes of gas hydrates were released, and this raised the earth's temperature by 5°C (Maslin 2004).

Another source of global warming is the variation in the brightness of the sun, already observed over several centuries. The sunlight intensity fell by 4 to 6% from the 1950s to the 1980s. Now it is rising again and there has been a 4% rise since 1990; it is estimated that this could account for a rise in temperature of 0.4°C by 2100. There are also daily variations of around 0.2%, and these could produce significant changes in the climate. In addition, there is a correlation between the temperature and the length of the sunspot cycle. The physical basis for this is suggested by another correlation, namely that between the cosmic ray intensity and the low cloud cover. The sunspot cycle is a measure of the solar activity, and this in turn affects the cosmic ray intensity. The cosmic rays produce ions in the atmosphere, and these can form condensation nuclei for clouds, which have a strong influence on the earth's temperature. Detailed studies suggest that solar effects may be responsible for 30 to 57% of the observed global warming. Since this varies with the sunspot cycle, it sometimes reinforces and sometimes weakens the effects of global warming. Some fluctuations have been observed and these could be due to varying amounts of aerosols in the atmosphere (Lomborg, op. cit. pp. 276–278). Recent work by Sloan and Wolfendale has however shown that there is no correlation between the cosmic ray intensity and cloud formation.

7.5. The Effects of Global Warming and Climate Change

There are some beneficial effects of global warming such as the extension toward the poles of the area of agricultural land in Canada and Northern Europe, and possibly New Zealand. Most of the effects however range from inconvenience to disaster. They include extreme weather conditions, melting of the polar ice caps and glaciers, and disruption or extinction of flora and fauna.

The Report by Sir Nicholas Stern in 2006 examined the current scientific consensus on the possible consequences of global warming and the economic costs. He detailed some of the likely consequences of different temperature rises:

1°C: Smaller mountain glaciers disappear in the Andes, threatening the water supply of 50 million people. More than 300,000 extra people die from increase in climate-related diseases in tropical regions.

2°C: Water scarcity increases in southern Africa and the Mediterranean. Significant decline in food production in Africa, where malaria affects up to 60 million people. Up to 10 million extra people affected by coastal flooding each year.

3°C: Serious droughts in southern Europe occur once every ten years. Between 1 and 4 billion people suffer water shortages and a similar number suffer from floods. Many millions of people are at risk of malnutrition, as agricultural yields at higher latitudes reach peak output. More than 100 million people are affected by the risk of coastal flooding.

4°C: Sub-Saharan Africa and the southern Mediterranean suffer between 30 and 50% decrease in availability of water. Agricultural yields decline by 15–35% in Africa. Crops fail in entire regions. Up to 80 million extra people are exposed to malaria.

5°C: There is a possible disappearance of the large glaciers in the Himalayas, affecting the supply of 25% of the population of China and hundreds of millions more in India. Ocean acidity increases with threat of total collapse in the global fisheries industry. Sea levels rise inexorably, inundating vast regions of Asia and about half of the world's major cities, including London, New York and Tokyo.'

This implies that the USA, Brazil, the Mediterranean countries, East Russia and the Middle East, Southern Africa and Australia are likely to lose up to 30% of their rainfall by 2050, and increase by a similar amount in Central Asia, Pakistan, India, Siberia and most of China. This will have a large effect of the people of India because the monsoon is very sensitive to rainfall: fluctuations of only 10% from the average value can lead to drought or flooding. The melting of the Himalayan glaciers could ultimately lead to extensive desertification. Many of the effects of global warming are very long-term, but are nevertheless devastating.

The conclusion of the Report is that if no action is taken the world could lose at least 5% and possibly up to 20% of global GDP each year indefinitely equivalent to £3.68 trillion. If, however, action is taken to reduce greenhouse gas emissions this could be limited to 1% of GDP each year, equivalent to £148 billion. This 'is an international problem requiring international co-operation and leadership' (Nuclear Issues 28, November 2006).

There are several other possible effects of climate change. There are many cases of communities that have flourished for centuries before being destroyed. One example is the Anasazi who lived in Colorado and were finally forced by major droughts in 1130–1180 and 1275–1299 to abandon their cities and to move away to areas with a climate that allowed them to continue their lifestyle. Greenland, as its name implies, was at one time a fertile land, and supported a colony of Vikings from about 1250 to 1650 when colder weather forced them to move elsewhere. The same cold spell around the fourteenth century caused parts of the Baltic Sea to freeze, and also the river Thames. In the nineteen thirties the reduction in rainfall on the Great Plains in the USA, followed by winds that removed the topsoil and created a dustbowl, forced farmers to pack up and move away. In other parts of the world weather patterns are subject to violent changes. The normally regular monsoons in India, for example, can sometimes fail, causing catastrophic famines.

One of these catastrophic famines occurred in 1899 and stimulated Sir Gilbert Walker, the Director General of Observatories in India, to study monsoons in more detail. He found that monsoons are not just a local phenomenon but are part of a vast global climate fluctuation called the Southern Oscillation. This consists of a cyclic process whereby high pressures in the Pacific are correlated with low pressures in the Indian ocean, and vice-versa. The period of this oscillation is about three or four years (Philander 1998).

Walker's work was recognised in the 1950s, when Bjerknes realised that the well-known El Nino current is part of the Southern Oscillation. He suggested that the Southern Oscillation is a periodic oscillation between El Nino and a complementary state called El Nina. El Nino has long been known as a warm ocean current that can have disastrous effects on the eastern Pacific shores. Satellite observations and computer modelling now enable some predictions of its effects to be made (Fagan 2004). Thus for example severe floods and storms were predicted to occur in California in 1997–1998. During autumn and over Christmas the weather was fine, but in January and February hurricane force winds battered San Francisco, floods rose and mudslides swept houses away. Floodwaters submerged the freeway to Los Angeles and swept away the Southern Pacific railroad bridge. Fourteen years earlier another El Nino caused floods and landslides that caused a billion dollars worth of damage. In other tropical regions, the 1997–1998 El Nino caused over $10 billion in damage. There were severe droughts in Australia and Southeast Asia, vast forest fires in Indonesia and Mexico and famine in Brazil.

The weather cycles in Peru and Bolivia are quite regular, unless they are interrupted by El Nino, which is unpredictably variable. Mostly the result is torrential rainstorms, warmer seas and changes in fish populations. Occasionally, however, an El Nino event causes significant changes to the climate and brings ruin to fishermen and farmers. Thus in 1925 the sea temperature off northern Peru rose over six degrees in ten days. Millions of seabirds perished as the anchovies on which they fed moved to cooler nutrient-rich waters. Cloudbursts turned dry ravines into raging torrents and the city of Trujillo received 396 mm of rain instead of the normal 1.7 mm. Farmlands and irrigation systems were destroyed by a sea of mud, and hundreds of people starved. Many other examples could be given of the devastating effects of El Nino.

It is conjectured that El Nino events played an important role in the decline and fall of ancient civilisations when they were already seriously stressed by other economic and political factors. Particular examples are the Maya civilisation in Yucatan and the Moche civilisation in Peru. In these cases the devastation caused by El Nino was the final event that caused the once-flourishing societies to collapse.

The rapid changes in climate associated with El Nino events took place long before the temperature increase associated with global warming and still continue. They are not yet fully understood and make it more difficult to ascribe climate changes to anthropogenic emissions.

Recent studies (Ker 2005; Webster *et al.* 2005; Witze 2006; Smith 2006) have shown greatly increased frequencies during the last 35 years for the more devastating hurricanes like the one called Katrina that struck New Orleans and the surrounding states in 2005. This increase has been attributed by some scientists to the rising temperatures of the oceans due to global warming (Witze 2006). If this is the case, some regions of the earth will be liable to more devastating hurricanes in the future. The cost of the damage due to the hurricane Katrina has been estimated to be around $100 billion.

On a much longer timescale, the mathematician Milutan Milankovitch identified cold and warm periods alternating every 100,000 years, with smaller cycles every 41,000 and 10,000 years. These cycles are attributed to perturbations of the earth's orbit by the moon that cause the precession of the equinoxes and by other small effects due to the planets and are confirmed by worldwide measurements of glaciers, coral reefs, peat bogs and polar ice caps.

There are thus many ways the climate can be changed in addition to the effects of the greenhouse gases. Careful scientific analysis is therefore needed before their contribution can be established. Many scientists worldwide are making detailed calculations using increasingly sophisticated models of the atmosphere. This is obviously a very complicated task. What, for example, do we mean by the temperature of the atmosphere? We can measure the temperature at a particular place and height, but this needs to be done over the whole surface of the earth and for heights up to several miles. The best we can do is to establish a grid of points and measure the temperatures at these points as a function of the time. Even a coarse grid contains millions of points and the calculations are very time-consuming even on a fast modern computer. The more accurate we want our calculations to be the longer they will take. In addition, the results may be very sensitive to the initial conditions; this is called the butterfly effect. The main uncertainty at present seems to be the effects of water vapour, which are greater than those of all the other gases combined. These are sensitively affected by changes in the cloud cover which in turn changes the amount of solar energy absorbed or reflected.

The results of such calculations are published periodically by the Intergovernmental Panel on Climate Change, consisting of about two thousands of the world's leading climate scientists, under the Chairmanship of Sir John Houghton. With many qualifications, the conclusion of the latest assessment (2007) is that there is unequivocal evidence that world temperature is increasing, and it is predicted that the average temperature will rise by about 0.2°C per decade. The sea level is predicted to rise by about 17 cm in the next century and this will eliminate many islands such as the Maldives in the Indian Ocean, and will inundate much of Bangladesh and some of Holland. Already the sea level has risen by 0.1 to 0.2 cm per year during the twentieth century.

The connection between the rise in temperature and the rise in sea level has been attributed to the melting of the polar ice caps. However, the ice immediately around the North Pole and in the ice shelves around Antarctica is floating, and so when it melts it has very little effect on the sea level, as Archimedes knew very well. There may, however, be some small effects due to differences in salinity between the ice and the sea.

Antarctica occupies 13.2 million sq.km, about 1.3 times the area of Europe and the ice cap is up to 4 km thick. No less than 90% of the world's ice is in Antarctica, and if this were all to melt the world sea level would rise by 70 to 90 m. However the ice in central Antarctica is at a temperature from –40 to –60°C and so is unaffected by a rise in temperature of a few degrees. The same applies to central Greenland which occupies an area about one-sixth of Antarctica. The ice shelves surrounding Antarctica are somewhat warmer, but are floating like the Arctic sea ice and so melting them has little effect on the sea level. However the sea level is affected by warmer coastal glaciers flowing into the sea from the ice caps of Antarctica, Greenland and Northern Canada. The glaciers are very thick, and the pressure on the ice where it rides over the ground is enough to liquefy it, and this makes it easier for it to slide down into the sea. An additional effect has been suggested: when the surface of the ice melts lakes are formed and this water flows down through cracks in the ice until it reaches the bedrock. There it spreads and reduces the friction between the ice and the bedrock, further increasing the rate of travel of the ice towards the sea. When this happens it causes the sea level to rise, but by an amount that is difficult to estimate. There is increasing evidence that the ice shelves are breaking up and this reduces the pressure on the glaciers and so their rate of flow increases (Shepherd 2006; Murray *et al.* 2008). The sea level rise is offset by increased evaporation from the sea, and so the net effect is difficult to determine.

The melting of the Arctic ice due to the rising temperature is already having a devastating effect on people living in the far north. In several places their traditional method of hunting seals is no longer possible because the ice has melted. The dry weather has caused widespread forest fires in Alaska, and the temperature of the permafrost has risen by two or three degrees. This leads to increased emission of methane, a greenhouse gas much more damaging than carbon dioxide. As the ice melts the albedo, a measure of the fraction of sunlight that is reflected falls rapidly from 0.8 to 0.9 to less than 0.1. The result is that more sunlight is absorbed, melting more ice in a continuing positive feedback effect. This is one of the main reasons why the Arctic ice is melting so rapidly, thus providing a sensitive indicator of global warming. Some computer models indicate that by 2080 the Arctic ocean will be ice-free in summer, making it impossible for the polar bears to survive (Comiso and Parkinson 2004). The Arctic ice has shrunk by more than a third since the late twentieth century, and the rate has accelerated faster than predicted by climate models.

Satellite observations have shown that the perennial Arctic sea ice covers about seventeen billion acres. This area varies from year to year, but in recent years the overall trend has been strongly downward, particularly in the Beaufort and Chukchi seas, and also to a lesser extent in the Siberian and Laptev seas. The shrinkage now amounts to about 250 million acres, and recent satellite and other observations show that the retreat is more rapid than estimated by any of the eighteen computer models used by the IPCC.

Many glaciers in Iceland and other mountainous countries are also retreating rapidly. Some of those in Iceland have retreated by as much as 1000 ft in a decade. Climate models predict that by the end of the century there will be ice only on the highest mountains.

The effects of global warming are not necessarily a smooth function of time. Thus according to one theory of Antarctic deglaciation at the start of the Holocene about 10,000 years ago a small change in temperature increased the temperature and hence raised the sea level. This destabilized the floating ice shelves which then broke up, exposing the margin of the ice sheet to attack by waves. This allowed the glaciers behind the ice shelves to flow faster into the sea, thus further raising the sea level. This process continued until feedback mechanisms eventually slowed it down. In this way a relatively small event can initiate a runaway series of events with significant consequences.

Another possible effect is due to the expansion of the oceans when they are heated. It is seldom mentioned that the land also expands when heated, and so whether this affects the sea level depends on the relative expansion coefficients of the land and the sea. This is further complicated by times that it takes for the warming to take place, which depends on the thermal conductivities of land and sea, and by the presence of currents in the sea.

Global warming is such a complicated problem, with many contributing causes and feedback loops, that there is still much debate about whether it is real or not. The climatologist Stephen H. Schneider has considered this question, and has pointed out that it depends on what we mean by proof. He considers that a strict deductive proof of the course of future events is impossible, but that 'my reading of the many lines of evidence puts global warming well over the minimum threshold of belief — far enough to assert that it is already proved to the point where we need to consider it seriously' (Brockman 2005, p. 76).

A recent report from the Environmental Protection Agency in the United States has highlighted the health effects of global warming, in particular concerning the food, energy and water on which society depends. They suggest that 'extreme weather events and diseases carried by ticks could kill more people as temperatures rise and allergies could worsen because climate change could produce more pollen. Smog, a leading cause of respiratory illness and lung diseases could become more severe'. These effects are already affecting human health and welfare.

7.6. Catastrophic Climate Change

The simple arguments concerning the connection between rise in world temperature and rise in sea level are inadequate, and so one often has to rely on the complicated world climate models. It is not possible for anyone who is not actually making such calculations to assess the validity or the results and, one may add, not so easy even for those who are. There is the additional uncertainty due to the possibility that the changes in average world climate do not always take place smoothly. This happens when some variable reaches a critical value and a large and irreversible change takes place, as when a kettle boils over or a house of cards collapses. We cannot be absolutely sure what will happen to the climate in the future. Thus it has been suggested that there may be a sudden and catastrophic cooling of northern Europe due to a change in the flow of the Gulf Stream. At present the Gulf Stream brings warm water from the tropics toward Europe by what is called the thermohaline circulation. This is due to the freezing of the Arctic water, which causes the salt water to drain out of the ice. This salty water is heavier than fresh water and so it sinks thus drawing warmer water northwards from the tropics. As this water cools it becomes denser and also sinks, thus attracting more warm water. If the oceans are heated by global warming and more freshwater enters the polar seas it could slow and even stop the Gulf Stream. This could cause the temperature to fall by six to eight degrees centigrade and so it would then be frozen for much of the year, and London would be like Siberia.

Such uncertainties are not uncommon in human affairs. We have to make a decision on the basis of incomplete knowledge. It is easy to say that we must undertake more research and do nothing until we are absolutely sure what is the best thing to do. This is nearly always the worst decision of all. We must take our decision on the basis of the best knowledge we have, even if it is to some extent uncertain. Concerning climate change, the best knowledge we have is contained in the results of the model calculations, and of how the climate has changed in the past from measurements on stratigraphic records from ice, lake, peat bog and marine sediment cores.

It has been estimated by the World Health Organisation that climate change is already causing about 150,000 deaths each year, and this is set to rise in the future. It is therefore prudent to explore any practicable means to reduce the emission of the gases that are responsible.

During the last century carbon emissions have fallen due to the gradual shift from coal to oil, gas and nuclear power, but soon this is likely due to change due to increased coal burning in China and India and reduced use of oil and gas. Governments are increasingly concerned about the effects of climate change, and have convened a series of conferences at Rio (1992), Kyoto (1997) and Johannesburg (2004) to decide the best ways to tackle the problem. At the meeting at Rio the representatives of 165 countries agreed to 'stabilise greenhouse gas emissions at a level that would prevent dangerous anthropogenic interference with the climate system'. As a start, countries were told to reduce their greenhouse gas emission to the 1990 level. By 1995, however, the emissions were still rising. At the meeting in Kyoto, the industrialised countries accepted rather different obligations: the Europeans to 8% below 1990 levels, the USA to 7% below and Japan to 6% below. This applies to five greenhouse gases in addition to carbon dioxide. The emerging industrial countries, including India, China, Sri Lanka, and Kuwait had no obligation to reduce in the period up to 2010. The USA refused to accept these targets unless they were applied to other countries as well. Otherwise, because reductions are expensive, the US companies would be at a disadvantage compared with companies in India and China. This is true, but it seems unjust for nations who have developed their industries without pollution control to prevent developing countries from behaving in the same way. At present the average American is responsible for the same emissions as 18 Indians and 99 Bangladeshi, so why should they have different targets? Despite strong international pressure the USA declined to take action, and its emissions rose by 12% during the 1990s. The British Government agreed to reduce carbon emission by 20% from the 1990 level by 2010, but it is evident that it is nearly impossible to achieve this (Nuclear Issues 28, 2006).

As already mentioned in Section 6.3, the potentially devastating effects of climate change require much larger reductions in carbon dioxide emissions, and it is clear that any realistic proposals, such as much higher taxes on emissions, would meet with implacable opposition. To overcome this is one of the most challenging problems of the present time.

7.7. Conclusion

The accounts in the last three chapters of the safety of alternative methods of energy generation, global warming and other environmental effects show once again a very familiar pattern. First there is the detection of some unexpected effect by a scientist. If this effect poses a threat to a well-established industry the industry tries to discredit the scientist's work, aided by other scientists who obtain different results. The journalists pay little attention, apart from writing articles about the general unreliability of scientists and The Scare that Never Was. Then another scientist obtains result that confirms the effect beyond any doubt, and shows that it could have seriously damaging effects. Immediately the journalists spring into action, blasting the wicked capitalists who are callously endangering the public and exaggerating the harmful effects. The scientist who made the original discovery may at this stage try to bring about a more balanced assessment of the situation, but he is completely ignored. Alarmed by all this, the Government or the United Nations sets up a high-level commission to investigate the matter. Meanwhile the industry that has its profitability threatened by the discovery redoubles its efforts to show that its products are quite harmless, and the journalists step up their campaign. At some stage the public become alarmed by the exaggerated accounts of the journalists and over-reacts by changing their lifestyle, even if the result is that they adopt an even more dangerous life-style. They also demand immediate Government action. The Government responds by assuring everyone that there is no immediate danger, and that nothing can be done until the Commission has reported its findings. The journalists, having stirred up the emotional storm, have now lost interest and are off chasing another scare story. The Government suddenly becomes convinced of the correctness of the discovery and realises that it must take action, whether or not that action is really effective, and whether or not there are much more useful ways of spending the money required. It knows that it will lose votes if it does not follow the mood of the public, even if by now it realises that it forces it to act in a way that are against the long-term interests of the people.

In such a situation, the leaders of the Church have an opportunity and a duty, namely to proclaim the truth, and to emphasise that action must be taken in accord with the long-term needs of humanity. Church leaders are not constrained by short-term political pressures or by the need to court popularity and they do not have to seek votes in order to remain in office. Speaking the truth may and probably will be unpopular, but this is not surprising. Indeed, their vocation as followers of Christ is not to be popular but to speak the truth.

References

BNFL Report and Accounts (2001).

Brockman, John (ed) (2005) *What We Believe but Cannot Prove* (London: Free Press).

Brown, Bob (2004) *Memo for a Saner World* (London: Penguin Books).

Carson, Rachel (1962) *Silent Spring* (London: Penguin Books).

Christianson, Gale E. (1999) *Greenhouse: The 200-Year Story of Global Warming* (New York: Walker and Company).

Comiso, Josefino C., and Parkinson, Claire L. (2004) Satellite-observed changes in the arctic, *Physics Today* (August), p. 38.

Diamond, Jared (1999) *Guns, Germs and Steel* (New York: W.W. Norton and Co).

DiMento, Joseph F.C., and Doughman, Pamela (eds) (2008), *Climate Change: What it means for us, our children and our grandchildren* (London: MIT Press).

Fagan, Brian (1999) *Floods, Famines and Emperors: El Nino and the Fate of Civilisations* (New York: Basic Books).

Fagan, Brian (2004) *The Long Summer: How Climate Changed Civilisations* (London: Granta Publications).

Hodgson, P.E. (1997) *Energy and Environment* (London: Bowerdean Publishing Co. Ltd).

Houghton, J. (1994) *Global Warming: The Complete Briefing* (Oxford: Lion Publishing).

Ker, Richard A. (2005) Is Katrina a Harbinger of still more powerful hurricanes? *Science* **309**, p. 1807.

Kolbert, Elizabeth (2005) *Meltdown* (London: Telegraph Magazine).

Kolbert, Elizabeth (2008) *Field Notes from a Catastrophe: Climate Change — Is Time Running Out?* (London: Bloomsbury).

Lomborg, Bjorn (2004) *The Skeptical Environmentalist* (Cambridge: Cambridge University Press).

Markham, S.F. (1947) *Climate and the Energy of Nations* (London: Oxford University Press).

Maslin, Mark (2004) *Global Warming* (Oxford: Oxford University Press).

Murray, T., Rutt, I., and Vaughan, D. (2008) The heat is on (Institute of Physics), *Physics World* (May).

Officer, Charles, and Page, Jake (1993) *Tales of the Earth: Paroxysms and Perturbations of the Blue Planet* [Quoted by John Carey in The Faber Book of Science (1995)]. (London: Faber and Faber), p. 355.

Philander, S. George (1998) *Is the Temperature Rising? The Uncertain Science of Global Warming* (Princeton: Princeton University Press).

Prance, Sir Ghillean (1996) *The Earth under Threat: A Christian Perspective* (Wild Goose Publications: Saint Andrew Press).

Shepherd, Andrew (2006) Antarctic unravelled, *Physics World* (May), p. 21.

Sloan, T., and Wolfendale, A. (2008) Physicists say clouds are not linked to cosmic rays (Institute of Physics), *Interactions* (May).

Smith, R. (2006) Hurricane force, *Physics World* (June) p. 32.

Vecchi, G.A., *et al.* (2006) Weakening of tropical pacific atmospheric circulation due to anthropomorphic forcing, *Nature* **441**, p. 73.

Webster, P.J., Holland, G.J., Curry, J.A., and Chang, H.-R. (2005) Changes in tropical cyclone number, duration and intensity in a warming environment, *Science* **309**, p. 1844.

Witze, Alexander (2006) Bad weather observed, *Nature* **441**, p. 564.

Chapter 8

Politics, Psychology and Education

8.1. Introduction

The previous chapters have shown that many of the technological problems of our society have become so distorted by political and psychological factors that there is a serious danger that the wrong decisions will be taken, imperilling our future. These matters are considered in more detail in this chapter, and connected moral decisions in Chapter 10. The possibility of a long-term improvement depends on education, and this is also considered in the present chapter.

During the last few centuries the world has become increasingly globalised. The financial systems and the markets for a wide range of goods are no longer limited to particular areas, countries or continents; they extend over the whole word. This has many advantages and disadvantages. It has raised the standard of living of millions of people, and provided food and manufactured goods to all countries much more efficiently and cheaply than ever before. There are also serious disadvantages: a financial crisis in one country affects many others, often causing acute distress. Flourishing industries are rendered unprofitable or even bankrupt overnight due to sudden changes in the exchange rate between countries. Globalisation has made millions of people worse off, as their jobs are lost and their lives become insecure. They feel powerless and in the grip of forces beyond their control. To prevent such disasters, international financial authorities have been created, but all too often they have been more concerned to safeguard the profits of the financiers than to address the worldwide problems of poverty and pollution. Battles are fought over the merits and disadvantages of government ownership versus privatisation. There are strong forces in favour of reform and if in future wise decisions are taken, gobalisation can still be a force for good (Stiglitz 2002).

The basic problem of the present time is that there are more and more people on the earth and they expect to enjoy the high standards of living made possible by technology, whereas the earth has insufficient resources to satisfy them. As a result of medical advances people, especially in the developed countries, are on the whole far healthier than at any time in the past. Indeed, most of us would not be alive at all without modern medicine. Infant mortality has been greatly reduced, killer diseases eliminated and the expectation of life greatly increased. Taken together, this has led to a rapid increase in world population, the plundering of the resources of the earth, pollution and destruction of the environment. In the poorer countries, where the population is increasing most rapidly, this leads to deforestation, desertification and a downward spiral of famine, disease and starvation. Many such countries lack the resources, the people with specialised knowledge and responsible leadership abilities to tackle this situation. Outside aid is suspected of neo-colonialism or economic imperialism and even when it is accepted it can easily be diverted into the private bank accounts of the leaders or the middlemen, and never reaches the poor people for whom it is intended. Thus the devoted effort of scientists, technologists, engineers and medical doctors to improve the lives of the people is frustrated by politics, greed and inefficiency.

We all like to believe that all our actions have the highest motives: we work to support our family, help others when they are in need and perhaps in our spare time undertake some charitable work for the good of mankind. This is indeed very often quite true, but sometimes other motives are at work, whether we realise them or not. To support our family we need money, and perhaps we obtain our money in somewhat questionable ways. If we are responsible for a factory, we may produce goods cheaply, so that they sell well, but we endanger our workers by neglecting safety standards and pay them the lowest possible wages. We may justify this by referring to our duty to the shareholders, but really the only thing we really care about is profit. If we are a journalist, we know that our stories will be published if they are full of sensation, sex and scandal. We invade people's private lives to extract stories, irrespective of the distress we cause. We may start our career intending to keep to the truth but gradually we are subjected to such pressures that we end up by being concerned mainly for what sells. We may give money to charity, but our motive is to draw attention away from our other activities and establish a reputation as a worthy citizen.

Mixed and hidden motives are so widespread that it is always necessary to look for them. Events can look quite different from different points of view (Pears 1998). The same applies on a local and national stage to politicians, captains of industry and indeed everyone with responsibilities beyond their own family.

It should be remarked that, as in many other sections of this book, the treatment is inevitably general, and deals mainly with ideas and not specific instances. As in all historical situations, the reality is infinitely complex. The situation differs from one country to another, from one university to another, and evolves with time.

The battle is between rhetoric and truth. We may recall the words of Newman: 'Quarry the granite rock with a razor, moor the vessel with a thread of silk, and then you may, with such delicate instruments as human knowledge and human reason, contend against those giants, the passion and the pride of man'.

8.2. The Problems of Democratic Societies

The increasing sophistication of our technologically-dominated society raises many problems that are likely to affects millions of people. For example, shall we welcome genetically-modified foods? Should we rely on wind-power for our future energy supplies? Should we forbid smoking because it causes lung cancer? How can we obtain the best answers to these and many other questions? We can try to do this democratically, by a popular vote, or we can ask experts to decide for us.

Both these methods are subject to serious objections. Very few people have the detailed technical knowledge to enable them to give a reasonably well-informed answer, and so the democratic method is an attempt to attain truth by summing ignorance. It is frequently said that scientists should inform the public; this is of course very desirable, but it remains very difficult to convey sufficient knowledge for the public to make an informed decision.

To place the decision in the hands of experts is much better, but there is still the problem of choosing the best experts. Political pressures with the object of putting the decision in the hands of people who are likely to say what you want them to say are very likely to confuse the process.

It is difficult enough to reach a sensible decision when everyone concerned is activated by the highest motives and concerned only to reach the best possible decision. The situation in our society is made much worse by powerful pressure groups activated by commercial or ideological concerns who orchestrate propaganda campaigns to support the decisions favourable to their aims. Thus tobacco companies devoted vast resources to show that smoking is harmless, and environmental groups support wind and solar power. Animal rights movements use violent methods to prevent research that is necessary to improve the treatment of diseases.

Can we rely on Governments to take the right decisions for us? They have access to the best scientific and technological information and thus are very well equipped to make wise decisions. It would be much easier to leave the decisions to the Government, but unfortunately this does not guarantee the best decisions. This is because they are very sensitive to public opinion, and this is strongly influenced by the commercial and political pressures just mentioned. If they go against public opinion, however misguided it may be, they will lose the next election. As a result, they are often under intense pressure to take decisions that they know are not the best ones. A familiar strategy in such circumstances is to postpone the problem by proposing a public enquiry. This may look prudent or statesmanlike, but it is disastrous in the long run. Very often postponing a vital decision is the worst option of all.

As long ago as 2000 a Royal Commission on Environmental Pollution said that Britain would need fifty new nuclear power stations to generate the energy needed without adding to global warming. In response to this recommendation Mr Blair commissioned a review of Britain's energy needs for the next fifty years. He declared that 'The aim of the review will be to set out the objectives of energy policy and to develop a strategy that ensures that current policy commitments are consistent with long-term goals'. It was confidently expected that this would lead to a revival of nuclear power but eight years later practically nothing has been done.

A democratic vote is useful only if the voters have access to expert technical knowledge. It is unrealistic to expect everyone to become experts, but at least they should have the opportunity to listen to the arguments of the experts. An expert is someone with the necessary basic scientific training who has studied at least some aspects of the problem for several years or, better still, has been engaged in research for some decades. It is not too difficult to identify an expert in this sense from readily-obtainable information. Such people will be the first to admit the deficiencies of their knowledge, since they understand the complexities of the problem. They welcome questions from people who genuinely want to know. It is usually the people who know nothing who are the most dogmatically sure that they know the best answer. If debates between experts are widely publicised by the media they could hopefully affect public opinion and lead to better decisions being taken.

Basically it is a matter of trust. We trust the airlines whenever we fly, we trust the shops to sell us food without harmful substances and so on. But an article in the Economist (10.07.2007) said that 'the polls reveal a striking and pervasive public distrust of official information about nuclear power. Only scientists employed by universities, a few television-news programmes and environmental groups are trusted to tell the truth. Government scientists and cabinet Ministers are widely disbelieved'. This statement is over-optimistic, and it would be unwise to trust all television programmes and environmental groups. University scientists are not always trusted. My own experience is limited, but I recall giving a talk on the energy crisis and nuclear power to some children in a State school. They sat in sullen silence during the talk, and afterwards I enquired how they found the talk and was told: 'They did not believe a word of it'. On another occasion I gave essentially the same lecture at Winchester College. The boys had already spent the whole day in the classroom and attendance at the lecture was voluntary. The room was packed and they were bright and enthusiastic, bubbling over with questions on what I was saying and what they already anticipated that I would say next.

8.3. Nuclear Power and World Peace

It is sometimes argued against nuclear power and nuclear weapons that it poses such a serious threat to world peace that all nuclear activities, including nuclear power plants, should be abandoned. However, nuclear weapons certainly pose a serious danger, but there are already several nations with nuclear weapons and it is too late to abolish them; the cat has already been let out of the bag. Furthermore, a major source of international conflict is the scramble for the remaining oil, which has already destabilised the Middle East.

Although the United States has been a leading oil producer since the mid-nineteenth century, its production peaked in 1972 and since then the rising demand has been met by importing oil. American companies have obtained oil from Saudi Arabia since 1939 and subsequently it has been a major supplier of US oil. As President Roosevelt said in 1947; 'the defence of Saudi Arabia is vital to the defence of the US'. US military bases such as that at Dharahan were established in the Middle East and in 1957 President Eisenhower was willing to use US troops to defend the area against Soviet Russia (Nuclear Issues, April 2005).

This policy was continued during the following years by President Nixon who supported Iran until the fall of the Shah. President Carter re-affirmed this policy, 'declaring in 1980 that to protect its vital interests the US would use any means necessary, including military force'. President Reagan increased the US forces in the Gulf and supported the Saudi royal family: 'There is no way we would stand by and see Saudi Arabia taken over by anyone who would shut off the oil'. When Sadam Hussein invaded Kuwait US and British troops repelled him. Subsequently Iraq was invaded to remove Saddam Hussein and secure continuing access to oil

If the USA considers it necessary to take such drastic measures to secure its oil supplies, it is optimistic to expect other powerful countries not to do the same. This is a potentially explosive situation.

The pressure to secure the diminishing oil is likely to intensify when the demand outstrips the supply, and this time is not far off. Conflicts that have already caused thousands of casualties and inflicted widespread devastation are likely to intensify and spread to other countries.

These very real dangers can be avoided, or at least rendered less severe, if another source of power becomes available, and nuclear power is the only source able to do this on the scale needed. Thus far from being a threat to world peace, nuclear power is the only way to avoid the escalation of the conflicts that are already taking place.

The existing stocks of plutonium, amounting to several hundred tons, can be used as fuel in nuclear reactors, the modern equivalent to beating swords into ploughshares. In 2004, 225 tonnes of weapons-grade highly enriched uranium from 9000 warheads was bought from Russia by the US Enrichment Corporation. When diluted, this uranium can also be used a fuel in nuclear reactors. The Russians plan to use the payment of $450M to improve safety, care of the environment and increased security.

As mentioned in Section 2.3, world oil production is likely to peak in a decade or two, and thereafter fall steadily. The supply of natural gas is expected to peak a decade later, and thereafter to fall rapidly. At the same time the demand for oil and gas continues to increase, so it is inevitable that soon the demand will exceed the supply. The price of oil will rise sharply and this will have serious and extensive repercussions. The most immediate effect will be on all forms of transport, which already use about 65% of world oil. The cost of transport will increase and many people will no longer be able to afford a car. Road transport of food and manufactured goods will become more expensive, and this will be passed on to the consumer. Already there is strong pressure to reduce carbon dioxide emissions from cars and lorries and this can be met by transferring to hydrogen-driven vehicles. The hydrogen has to be produced, and this can best be done without further carbon dioxide emissions by using nuclear power. The requisite technology is not fully developed, so it is likely that the first commercial hydrogen-powered vehicles will be more expensive than those at present in use. To overcome this, new taxation and other methods may prove necessary but even then it is very likely that the number of cars on the road will suffer a sharp decline. This in turn would cause vast social changes, as people who live in rural areas and work in nearby cities find it economically impossible to continue their lifestyle. This is particularly acute in large countries such as the USA and Australia. In Australia, for example, there are 13.2 million vehicles travelling an average of 15,300 km/yr and this uses 80% of the oil, with a further 10% for aviation. The situation of the USA with 210 million cars and lorries is similar. Even in Britain, with shorter distances and a well-developed public transport system, road transport still accounts for 62% of petroleum, and a further 20% for aviation. With the ending of the supply of cheap oil, the economic expansion of the last two hundred years will end. Many financial institutions are based on the assumption of continued economic growth, lending money they do not have in the expectation that it will be repaid by an expanding economy, so that if growth ceases they could face economic collapse, with immense repercussions. Poorer countries will have to pay more for their oil, putting a severe additional strain on their already overstressed economies.

The only practicable way to prevent this and to ease the painful transition from an oil-based economy to one based on other energy sources is a massive expansion of nuclear power. Even if this is done, the transition will inevitably prove to be a painful and socially-disruptive process.

To tackle the immediate situation concerning oil, the Association for the Study of Peak Oil has proposed the so-called Rimini Protocol which requires countries to restrict their oil production to their current depletion rate 'defined as the annual production as a percentage of the estimated annual left to produce and that each importing country shall reduce its imports to match the current World Depletion Rate, deducting any indigenous productions' (Nuclear Issues, June 2005). Such an agreement will encounter severe difficulties, but it does offer a possible way to ease the transition.

8.4. Energy Sustainability

A sound energy policy must ensure that the quality of life of the world's people should continue and even improve on the present situation, and that people below this level should be brought up to the level enjoyed in the developing countries. This requires not only the production of sufficient energy but also taking care that the consequences of that energy production do not destroy the amenities of the earth by pollution and climate change. This is what is meant by a sustainable energy policy. Energy production is considered in this section and the problems of pollution and climate change have been discussed in previous chapters.

In order to see whether we have enough energy to supply our future needs it is necessary to evaluate the capacities of the various energy sources and whether they will continue to produce energy in the future. Energy from inexhaustible sources are sustainable, while those that depend on finite amounts of raw material are not. The renewable energy sources considered in Chapter 3 are sustainable, whereas the fossil fuels and nuclear are not. As we cannot foresee the scientific and technological advances that will certainly be made during the present century, it is unrealistic to try to evaluate the situation beyond fifty years from the present. This re-defines sustainability as referring to the next fifty years and not to the infinite future. On this criterion both coal and nuclear become sustainable. This restriction to the next fifty years does not apply to the problems of pollution and climate change.

To see if our lifestyle is sustainable we need to calculate the average total energy needed or habitually used by each person and compare it with that obtainable from sustainable sources. This has been done with great care by MacKay (2008) and the following summarises some of his results. The analysis refers to Britain, as most of the necessary statistics are available; much of it applies to other developed countries. He does not consider the economics of the various sources as he wants to establish the absolute upper limits to the availability of sustainable energy. In practice economic considerations impose severe constraints that greater reduce the available energy.

Considering the energy expenditures in turn, he finds for each person that travel by car uses about 40 kWh/day and travel by air by about 35 kWh/day for each long-distance return flight and about 6 kWh/day for a return short-haul flight. Domestic use averages 14 kWh/day, including cooking, cleaning (washing machines and dishwashers) and cooling (refrigerators and air conditioners). If electric fires are used for heating this uses 24 kWh/day, giving a total domestic use of 38 kWh/day. The lighting of homes, factories and streets requires about 4 kWh/day. Other household appliances (computers, radios, television and phones) use 5 kWh/day, food 12 kWh/day, commercial transport 12 kWh/day and other miscellaneous uses 10 kWh/day, a total of about 80 kWh/day.

The corresponding analysis of the maximum available energy from different sources requires a number of assumptions that can be questioned, but these are chosen deliberately to be on the optimistic side. MacKay then sketches six scenarios with different contributions from the various sources, insisting that they add up to the required total. All but one include nuclear power or coal. The remaining one relies heavily on wind power and requires a hundred lakes or lochs for energy storage. Thus even from the point of view of energy sustainability, without reference to costs, it is hard to see how the renewable sources can meet our needs.

8.5. The Opposition to Nuclear Power

We can begin by asking why there is such sustained and vociferous opposition to nuclear power. This campaign has now convinced so many people that nuclear power is unacceptable as an energy source that Governments refuse to build nuclear power stations. If they seem to favour nuclear power in any way, they know that they will lose votes, and that is what Governments want to avoid at all costs. The arguments in favour of nuclear power are now so strong that this all seems incomprehensible.

It is sometimes suggested that the origin of the opposition to nuclear power is the fear of atomic bombs. This is not so. In the immediate post-war years many scientists devoted considerable efforts to educate the public: they wrote articles and books and organised lectures and exhibitions. They founded the Federation of Atomic Scientists in the USA and the Atomic Scientists' Association in Britain to further this work. The result of these efforts was general public enthusiasm for the potentialities of nuclear power, and people looked forward to the Atomic Age.

This gradually changed as a result of a well-planned and large-scale campaign by the Soviet Union designed to weaken the West. It was clear to the Soviet leaders, whose aim was world domination, that the economy of the West depends on the availability of energy, so they first tried to jeopardise the supplies of oil by destabilising the countries in the Middle East that supply so much of Europe's oil. It then became apparent that the West could obtain much of the needed energy by developing nuclear power, and so this became a prime target. A massive propaganda campaign was therefore launched against nuclear power. Of course, like all well-designed campaigns, it pretended to act for the public good. It was based on genuine science, but distorted to serve its propose. The dangers of radiation were correctly described, but grossly inflated; it was not pointed out that with proper care, these dangers are negligible. Nuclear reactors were described as likely to blow up and spread radioactivity everywhere; it was not pointed out that reactors in the West were so designed and managed that this could not happen. (The accidents at Three Mile Island and especially that at Chernobyl were a godsend to the campaign.) The disposal of nuclear waste was declared to be a vast and unsolved problem. People were horrified by these revelations, and took no notice of the explanations of the nuclear scientists. The campaign was cleverly designed to enlist the support of well-meaning people who lacked the technical knowledge to understand how they were being deceived. It was spearheaded by Communist parties throughout Europe and taken up by left-wing parties everywhere. Finance on a massive scale to support the campaign was provided by the Soviet Union (Andrew and Mitrokhin 1999). After the collapse of the Soviet Union, many hard-core leftists found themselves without a cause, and so transferred their energies to the anti-nuclear campaign.

Emotion and rhetoric proved once again more powerful than scientific analyses and statistical data. People joined the campaign in their thousands, went on protest marches and contributed their money. Left-wing politicians and academics joined the chorus of condemnation. Gradually the mood of the public changed, and the enthusiasm and euphoria of the sixties was replaced by determined opposition. Especially after Chernobyl, country after country cancelled their nuclear power construction programmes and abandoned nuclear power. The Swedish Government closed its Barseback 1 reactor after it had operated for 23 years, and the resulting shortfall of 4 billions kWh per year had to be obtained by importing it from German and Danish coal power stations, thus increasing Sweden's carbon dioxide emissions (Nuclear Issues 23, March 2001). Subsequently the Barseback 2 reactor was closed as well, although it still had 20 to 30 years of productive life remaining. Again, for political reasons Germany dismantled a 1300 MWe nuclear power plant at Mulheim-Karlich that had only operated for two months.

While attacking the Western nuclear power programme, the Soviet Union feverishly built their own reactors, designed to produce weapons grade plutonium as well as power. Their reactors were so poorly designed, so hurriedly built and so poorly operated that the disaster at Chernobyl was almost inevitable.

In this way politics triumphed over truth.

One might have hoped that those who control the mass media would have understood the origin of the campaign against nuclear power, and have considered it their duty to print and broadcast the truth. That would be a vain hope. Newspapers are engaged in a life and death struggle with their rivals, and television producers are under huge pressure to maintain and increase their ratings. They know very well that their readers and viewers are attracted by sensational scare stories, whereas the comments of the scientists, full of jargon and incomprehensible numbers, are certain turn-offs. So if someone says that there are far more cases of childhood leukaemia among people living near nuclear power stations this is given blazing headlines. Any subsequent analysis by scientists, explaining carefully that the increased radiation around power stations is minuscule and cannot possibly cause such effects is simply ignored. It is seen as an attempt at a cover-up by the nuclear industry. Such stories have encouraged some families with affected children to take the nuclear industry to court. The medical evidence is such that they have lost their cases.

When a scare story breaks, there are always anti-nuclear activists, with little knowledge and no reputation to lose, who are willing to make wild and unsupported statements that are eagerly quoted by the media. Genuine scientists are not willing to make statements until they have examined and weighed the evidence, and this may take weeks or months, by which time it is too late to say anything that will be heard.

Many arguments were used against nuclear power. One of the most plausible began by pointing out, quite rightly, that there are still many dangers and uncertainties associated with it. Would it not therefore be prudent to declare a moratorium on future development until it can be shown to be perfectly safe? It is such a promising source of energy that it is important to avoid undue haste that might do great harm. What could be more prudent and statesmanlike to wait for five years and see if by then it has been made perfectly safe? This is a very attractive idea, but it is simply a delaying tactic masquerading as concern for humanity. Many people, including Church leaders and the World Council of Churches, fell into this trap. If this argument had been accepted, then after five years it would be repeated again and a further moratorium proposed. The opponents of nuclear power would never have accepted that it had finally been proved safe. By that time the technological expertise would have dwindled and huge financial losses incurred, and it could then be argued that nuclear power is uneconomic. It has to be recognised that no technological process is perfectly safe, so what has to be done is to examine all the ways of producing energy as critically as possible, taking into account other considerations such as cost, capacity, reliability and effects on the environment, and then make our choice. To do nothing is often the worst choice of all. It is greatly to the credit of the experts convened by the Pontifical Academy of Sciences that they accepted the need for decision and speedy action, and vehemently rejected the call of a moratorium (see Section 10.4).

Another campaign designed to discredit nuclear power that was disguised as a noble humanitarian gesture was to bring children from the region around Chernobyl to Britain and to provide them with medical attention and a healthy life. These children were designated 'victims of Chernobyl' who had been exposed to deadly nuclear radiations that would greatly shorten their lives. Pictures of the children and their tragic stories were given great publicity in the media. In fact there was no evidence that they had been exposed to abnormal amounts of radiation, and their generally poor state of health was due to decades of malnutrition.

The campaign against nuclear power has also used the clusters of leukaemia cases around Sellafield and the effects of the discharges of radioactive material into the sea. These questions have been discussed in Section 4.5 on nuclear radiations.

This campaign was continued for decades by the media and in schools and universities, and eventually emotion and rhetoric proved once again more powerful than scientific analyses and statistical data.

There is a hard core of people who are opposed to the modern technological society as such; they consider it to be evil and do all they can to destabilise it. They do not care whether the arguments they use are true or not. The same arguments against nuclear power are repeated again and again and are printed by the media as if they were a new revelation although they have been repeatedly refuted. They make no attempt to assess the advantages and disadvantages of any new technology such as nuclear power or genetic modification, and ignore the great advantages they can bring in the form of the energy and food so desperately needed by the poorer people of the world. All this is of no concern to them.

Strong opposition to nuclear power comes from members of the post-modernist anti-science movement. It is difficult to discuss anything with them because they deny the whole idea of objective truth, maintaining that all beliefs are subjective, and depend on personal whims, desires and prejudices. They do not accept that people believe anything because they have considered the arguments and reached a decision. Thus according to them some people accept nuclear power for some irrelevant reason such as their presumed support of the prevailing political-industrial-military complex. It is hard to reason with people who do not accept reason. Some people have so far lost the concept of truth that they assume that everyone acts for some ulterior motive. Thus I was once asked if I was paid by the Atomic Energy Authority to speak in favour of nuclear power.

Concern for the environment is rightly a very popular concern, and this has been used in the campaign against nuclear power. However the policies advocated by many environmentalists are often more destructive than the alternatives. This has led several distinguished people such as Bishop Montefiore to resign their membership of prominent environmentalist organisations. The environmental scientist James Lovelock, noted for his extensive researches that have led to the concept of gaia, the idea of the earth and its ecosystems as a closely interdependent organism, has also dissociated himself from extreme environmentalists and declared his support for nuclear power as the only way to save the environment.

It is sad that many environmentalists refuse even to discuss the advantages of nuclear power for the environment. I recently read a book by an environmentalist who described his excellent work in opposing unnecessary hydroelectric developments in Tasmania and the destruction of the forests by logging (Brown 2004). He appended the Green Party manifesto that declared its opposition to nuclear power. I wrote several times to him asking for the justification for this, but received no reply. Such people are so sure of the correctness of their views that no dialogue with them is possible.

After the Chernobyl disaster there were stories of increased numbers of deaths, even as far away as the USA, due to the radioactive cloud that spread over the northern hemisphere. In one newspaper, the story was illustrated by a grim picture of the death of the reaper. This story seemed very implausible: the doses in Europe were far too small to cause such effects, and indeed none were recorded, and those received in the USA were far smaller. So I obtained from the Atomic Energy Research Establishment, Harwell, the numerical data on which this story was based. It was then evident that the data did not support the story, due to an elementary statistical error. I wrote to the newspaper explaining this, but they were not interested in publishing a correction; they were pushing the latest scare story. Even if they had published a correction few people would have noticed it, and all the readers would be left with the image of death the reaper associated with nuclear power.

Disasters like Chernobyl have led to calculations of the number of cancer deaths to be expected from such emissions of radioactivity. The death rate attributable to large doses of radiation is known, and it is then assumed that the death rate at smaller doses can be obtained by assuming that the death rate is proportional to the dose. This gives a very small figure, but when this is multiplied by the population of a large country a figure of some thousand deaths is obtained. Of course the fallacy is to assume proportionality, which ignores the possibilities that there is a threshold dose and that the body can repair small damage (see Section 4.5). Some time ago a prominent left-wing politician attacked nuclear power, using just this argument. I wrote to him asking for justification, and he referred me to a source that he could not then remember, but which I could find for myself if I took the trouble, but he doubted if I would do so. I asked him if I could publish his remarks, and he indignantly refused. This is typical of what happens when a scientist tries to inject some truth into what is euphemistically called the debate on nuclear power.

On another occasion I wrote an article on the Sellafield reprocessing plant and among other things I pointed out that coal power stations actually emit more radioactivity that nuclear power stations, although in both cases the amounts are minuscule. The reason for radioactive emission from coal power stations is that coal contains very small amounts of uranium, and some of this is emitted as smoke into the atmosphere. Nevertheless my statement was considered so preposterous that it was rejected out of hand. So I wrote again giving the detailed statistics (see Table 6.5) based on measurements compiled by the International Atomic Energy Agency that supported my statement. This letter was not published.

These experiences show that the reactions of politicians, journalists and scientists to the news of a nuclear accident are quite different. Politicians will be concerned to assess the effect on their own reputation and that of their Party, and most importantly on their chance of re-election. They welcome the news if it seems to be to their advantage and will downplay if it does not. In general, they decide their policies so as to follow the contemporary beliefs about what should be done. If these really are for the general good, and often they are, that will be done. The problem arises when Governments see what needs to be done, but also know that it is very unpopular, and then they will procrastinate and follow the general will. By the time the effects make themselves felt, the next election will be long gone. If anyone protests, they will be ignored or brushed aside. The anti-nuclear activists among them will eagerly publicise any accident, and embellish their accounts with lurid tales of the consequences. They are absolutely sure that they are right, and if anyone so much as gently enquires about the logical reasoning behind these stories he is likely to be accused of impertinence and stupidity*.

One might hope that politicians would have the courage to take the hard decisions that are required to safeguard our long-term future. They have before them the evidence of countless scientific studies and reports by the Royal Society and the Federation of British Industry. However the next election looms larger in their minds than the future of mankind. The evidence for climate change is now so compelling that Governments have to be seen doing something about it. They dare not risk offending public opinion by choosing nuclear power. Instead they fasten on the renewables, especially wind power as a safe political choice, in spite of the arguments showing its futility. Conferences are arranged to consider the problem of global warming and climate change and consider all means of preventing these harmful emissions such as improved energy efficiency, carbon emission taxes and wind and solar power, except the one source that is demonstrably the only practicable way to solve the problem. Nuclear power is taboo. To further bolster their popularity, the Government seems determined to destroy our nuclear industry. There is now hardly any research and development of nuclear reactors, and flourishing nuclear enterprises like Bruce Power in Canada have been sold for derisory prices. When the pressure of events forces us to return to nuclear power, we will have to obtain the technology from France, Germany or Japan.

Journalists will welcome the news of nuclear accidents, and will go ahead and write scare stories, adding fuel to the flames by recalling similar accidents that happened in the past. They know that people love reading about disasters and that accounts of them sell their papers and increase their popularity. They have no interest whatsoever in knowing whether the story is true or not, and are quite indifferent to arguments that throw doubt on the report. If a scientist writes to correct these accounts, he is usually ignored.

The scientist will react to the news by asking himself if it is true. He will compare the reported events with what is already known about similar events in the past. Is the new event in accord with the laws of

---

* Several examples of this behaviour are given in Hodgson, P.E. (1999) *Nuclear Power Energy and the Environment* (London: Imperial College Press), pp. 161–165. See also *The St. Austin Review*, September/October 2004, p. 43–44.

nature? What are the results of studies of similar events? If it reports statistics about injuries, he will want to examine the methodology of the observations, how were the results evaluated statistically, and was there a control group for comparison? Is there any reason to suspect the results, does the laboratory have a good reputation for careful experimentation? Is it possible that in some way the political views of the scientists influenced their work in any way? This process may take months or years, but the scientist must keep an open mind before publishing his analysis. By the time he has made such a critical analysis, Congress or Parliament will have decided on their course of action and the conclusions of the scientist have had no effect. The media will no longer be interested; they are already describing the latest scare story. The scientist is of course appalled by all this, and may write a letter to the newspaper, but it will almost certainly be ignored. Soon he gets tired of writing letters that are never published, and so he writes no more.

It should also be mentioned that there are some writers, with some scientific training, who may describe themselves as nuclear consultants, who amass and distort scientific data in such a way as to support the anti-nuclear case. To the uninitiated it appears scientific and convincing, and only a real scientist can see how the data have been twisted, exaggerated or ignored in order to present their case. Their writings are of course used as a source of data and are widely quoted by journalists and all who oppose nuclear power.

Well-informed critiques of nuclear power have at least the merit of providing the opportunity to continue factual discussions. More damaging is the tendency of some writers to ignore nuclear power or to brush it aside with a few dismissive and often erroneous remarks. Thus Al Gore in his valuable and wide-ranging book 'Earth in the Balance' devotes barely a page to nuclear power. He notes 'the political difficulties involved in building new nuclear reactors' and states that 'uncertainties about future projections of energy demand and economic problems like cost overruns were the major cause of the cancellations by utilities, well before accidents like those at Three Mile Island and Chernobyl heightened public apprehension'. Predictably he raises again the long-laid spectre of nuclear waste, remarking that 'growing concern about our capacity to accept responsibility for storing nuclear waste products with extremely long lifetimes also adds to the resistance many feel to a dramatic increase in the use of nuclear power'.

Without giving any supporting statistics, he declares that 'the proportion of world energy use that could practicably be derived from nuclear power is fairly small and likely to remain so. It is a mistake, therefore, to argue that nuclear power holds the key to solving global warming'. The most cursory glance at the contribution of nuclear power to world electricity production shows that this is completely false. He does, however, admit that further research into new types of nuclear reactors, including fusion reactors, 'should continue vigorously' (Gore 2007, pp. 328–329).

Another example is provided by the valuable and well-informed book on 'Global Warming', where Mark Maslin just says that 'many countries are also discussing renewing their nuclear programme as a non-carbon-emission source, but problems of safety and dumping nuclear waste still remain the main objections' (Maslin 2004, p. 141). Two pages later, he concludes that 'ultimately, a combination of improved efficiency and alternative energy is the solution to global warming'.

The reactions of the Churches will be described in Chapter 10.

8.6. Psychology

In the present context psychology will be taken to mean the study of irrational behaviour: why do people behave in ways that are completely unjustifiable? Sometimes there are underlying reasons for this, but in other cases it remains obscure. There is obviously considerable overlap with the behaviour described in the previous section on politics.

Very often the conclusions reached on complicated technical problems depend more on preconceived attitudes than on objective scientific evidence. In most cases superficially plausible arguments can be made both for and against any proposed course of action. This primacy of attitudes over beliefs was already pointed out by Locke and Hume. Our attitudes determine the weight we give to arguments. Thus if we are averse to nuclear power we will tend to believe that leukaemia clusters around nuclear plants such as Sellafield are caused by nuclear radiations, that the safe disposal of nuclear waste is a great unsolved problem, that the Gardner hypothesis about the effects of parental irradiation is preferable to the explanation given by Kinlen, that all our energy shortage can be solved by using renewables and so on.

All these attitudes can be shown to be incorrect, but they nevertheless persist.

In a situation with many arguments and counterarguments there is a strong tendency for judgements to be affected by political and economic considerations. Thus those responsible for the fossil fuel generation will tend to downplay climate change, though it is more difficult to avoid admitting the pollution by other chemicals. The proponents of nuclear power, on the other hand, will emphasise climate change because this enhances the desirability of non-polluting nuclear power. Governments will be attracted by the idea of imposing taxes on carbon emission because it is a lucrative source of revenue.

Attitudes are reinforced by emotive words. Thus the disposal of nuclear waste is described as 'dumping', a word that conveys careless behaviour by people who ignore the likely results of their action. Some new energy sources are called the 'benign renewables' although they are not particularly benign and are not, strictly speaking, infinitely renewable. There is liberal use of soothing phrases like 'care for the environment' and of words like 'green' and 'peace'.

The primacy of attitudes over belief and the general gullibility of so many people is also shown by what Park (2000) has called pathological science, junk science, pseudo-science and fraudulent science. Taken together, they may be described as Voodoo science. Pathological science occurs when reputable scientists manage to deceive themselves. They observe some effect that seems at first to be a great discovery, but are unable to bring themselves to make the crucial experiments that would show whether it really is what they think it to be. Junk science is when scientists concoct plausible but scientifically false arguments in order to confuse or deceive juries and judges who lack sufficient knowledge of science to realise that they are being fooled. Examples of pseudo-science are types of spiritual healing based on quantum mechanics, reports of the Earth being visited from Mars and the whole range of quack medicines. Initially, the exponents of these views may really believe what they say, but if belief wanes and yet they continue their activities they gradually slip over into fraudulent science.

Pathological science typically begins when an experiment gives a result that is plausible at first sight but extremely unlikely on further examination. One has to be very careful at this point. When we say that a result is extremely unlikely we mean 'according to the present laws of physics', and of course we know very well that there have been genuine surprises in the course of scientific history. We know that there are laws that have stood the test of time, such as the conservation of energy, and so it is extremely unlikely that an exception can ever be found. Even in this case, it was later found that mass can be converted into energy and vice-versa, so that instead of the two laws of the conservation of mass and of energy we have the conservation of mass-energy. Another law that cannot be circumvented is the second law of thermodynamics.

In many cases our confidence in these laws is so strong that any device that purports to act in a way that goes contrary to them can be dismissed without detailed examination. Thus it was long ago decided that claims for any device that claims to produce more energy than it consumes, and for any alleged perpetual motion machine, can immediately be rejected. The same applies to the gravity shield, because if it were possible to shield any body from the effects of gravity it would be easy to construct a perpetual motion machine.

Experienced scientists who are familiar with a certain type of phenomena generally know what sort of result to expect from an experiment. Occasionally they are surprised, and then it usually turns out on further investigation that there is a perfectly intelligible reason for what they observed. The tip-top and the gyroscope can serve as simple examples.

It may happen that a new experiment with an unexpected result opens up vistas of practical applications and thus of money-making on a huge scale. This immediately generates psychological pressures that some find difficult to resist. In spite of all this, the scientist must curb his enthusiasm and make careful tests to verify the result. If the effect occurs, it will have such and such consequences, and these must all be verified. If all the tests are passed, the work is written up for publication. The apparatus must be described in such detail that it can be replicated elsewhere, so that other scientists can verify the result. The article will then be peer-reviewed and if accepted will be published in a recognised journal. Then, and only then, it is time to call in the Press and announce the result to the world.

A good example is provided by cold fusion. The story is so familiar that it is hardly necessary to describe it in detail. It is well-known that the heat of the sun comes from the fusion of hydrogen nuclei to form helium. In principle, the simplest way to do this is to take two deuterons, nuclei of deuterium, a heavy isotope of hydrogen (each comprising a neutron and a proton) and combine them to form an alpha-particle, the nucleus of helium. However, a simple calculation shows that the two deuterons repel each other electrostatically so strongly that it is just not possible to make them combine at room temperatures. They will only combine if they are heated to temperatures of some hundreds of millions of degrees, and physicists are trying to do this by designing fusion reactors. The great attraction of achieving fusion is that deuterium occurs in ordinary water, and so fusion could provide an essentially limitless source of energy.

Two electrochemists, Martin Fleischmann and Stanley Pons from the University of Utah, claimed in 1989 that fusion had been observed during electrolysis of heavy water (water containing deuterium instead of hydrogen) with a specially prepared deuterium-impregnated cathode of the metal palladium. This of course was a startling news. It had to be taken seriously because Fleischmann was a very distinguished scientist and professor at the university of Southampton, and Pons was a full professor of chemistry at Utah and both had a long list of publications. Scientists around the world rushed to make the experiment. Some, generally in the more poorly equipped laboratories, jumped on the bandwagon and announced success, but scientists in the major laboratories failed. Fleischmann explained that it was very difficult to make the palladium cathodes, but did not provide sufficient details. He also failed to make two crucial tests. Firstly, if fusion took place, helium should build up in the cathode, and its presence should be detected. Secondly, the helium should be formed in a highly excited state, and should decay by emitting neutrons. These neutrons should be detectable, but they were never conclusively found. In spite of these doubts, Congressmen argued that the potential rewards were so large that it was worthwhile providing massive support. Scientists were generally highly sceptical, and it was considered unacceptable that Fleischmann and Pons did not first publish their results in the usual way, with a full description of their apparatus, and so expose them to critical examination. Soon the whole affair died a natural death.

There is an instructive comparison to be made with the discovery of high-temperature superconductivity by George Bednorz and Karl Mueller in 1987. This effect was totally unexpected and was not predicted by existing theories. It also opened the way to potentially very important applications. The announcement caused great excitement and, like cold fusion, hundreds of scientists rushed to reproduce the result.

The crucial difference was that the discoverers had followed the rules and made a public announcement at the same time as a detailed account of their experiments was published in a peer-reviewed journal. Everyone who repeated the experiment obtained the same results.

A further comparison can be made with the alleged correlation between childhood leukaemia and proximity to nuclear power stations. The statistics, though rather poor, deserve to be taken seriously. It is natural to begin by asking whether the leukaemia is indeed caused by nuclear radiations. Calculations of the radiation doses that can be attributed to the nuclear reactors show levels that are far too small to cause the observed effects. Furthermore, examination of the distribution of leukaemia cases over the whole country shows several statistically significant clusters of cases in regions far away from nuclear power stations. The connection with nuclear power stations can therefore be ruled out, and the way is open to look for the real cause. This has been plausibly identified as the Kinlen effect, as already described.

This would have been a rational objective way of dealing with the matter. Instead, it was eagerly enrolled as part of the campaign against nuclear power, distraught parents of the afflicted children went though the trauma of making and losing claims against the nuclear power company, and a huge amount of money was wasted in vain attempts to prove the presence of any real connection.

Concern with their health leads many people to buy medicines that claim to cure their ache and pains. Among these may be listed bee pollen, vitamin O, homeopathy, biomagnetic therapy and a host of others. They are widely advertised and sold in huge quantities. Their success is supported by numerous testimonials from satisfied customers, but none has even been proved scientifically to have the effects claimed. The best that can be said for most of them that they do no harm, except to one's pocket. These are examples of fraudulent science.

An example of junk science is provided by the alleged effect of the electric fields from overhead power lines, in particular that they can cause cancer in people living near them. It is very easy for journalists to write tear-jerking stories about stricken children living near power lines, and these stories soon lead to the discovery of further cases. There is widespread concern long before a careful statistical analysis shows no evidence for the effect. Organisations of concerned people are formed to protest against the danger. This provided the occasion for what is known as a mass tort blitz. The idea is that lawyers encourage thousands of people to file lawsuits against the power company. The complaints are carefully worded. They do not say that electric fields cause cancer, because that can quite easily be disproved. Instead, they say that there was a widespread perception of hazard that has substantially reduced the value of their property. Thus they do not have to prove the existence of a hazard, but only that people believe that there is a hazard. No company can cope with such a situation and they are forced to come to a settlement. As it happened, judges realised what was happening and were able to invoke another law that put a stop to the whole campaign. This story provides an example of a totally unsubstantiated worry that was used by unscrupulous lawyers in an attempt to bankrupt a company. A careful scientific analysis of the strengths of the electric fields and the distribution of cancer cases would have soon shown that the fears were groundless.

Throughout history there have been people who believe in astrology. Our destiny, according to astrologers, is determined by which planets were in the ascendant when we are born, and thereafter our lives are governed by the stars. There is certainly evidence for the influence of the sun on our lives; it provides the heat and light without which we could not exist. The moon is the principal cause of the tides, with a lesser influence from the sun. These are well-understood effects of the radiation emitted by the sun and of gravitational forces. There the influence ends. The planets are much too far away to influence us either by radiation or by their gravitational attraction. The stars are at unimaginably greater distances. Yet although astrology is complete nonsense hundreds of books on astrology disgrace our bookshops and numerous magazines have an astrology page that purports to give advice on how to plan our week, based on the sign of the zodiac we were born under. Astrology is a very lucrative profession; there are more than ten times as many astrologers in the USA than there are astronomers.

Another familiar example of junk science is provided by all the stories of UFOs, little green men and other aliens from outer space. In one case, a farmer found some pieces of metal and plastic on his land, which were later shown to be the remnants of a weather balloon. He reported his findings to the local police and a journalist got to hear of it and suggested that they were the remains of a space ship. Then the story took off and with each retelling more details were added. The authorities had to take notice and ordered an enquiry. They were hampered by military secrecy, and soon it was being said that there was a massive cover-up to conceal the truth. And so it went on. Books were written and sold in large numbers. Eventually the official report, a massive tome, appeared. But few people were interested. It was much more interesting to read about the little green men.

Much more serious is the effect of military secrecy. No one in their senses will want to deny that there is a need for secrecy in matters concerning national security, but secrecy is not without its dangers. Proposals for new devices may be accepted without proper scientific scrutiny, and this can lead to an enormous waste of money. A spectacular example of this is the 'Star Wars' initiative by the USA. The President announced that he had ordered the construction of a nuclear shield, batteries of nuclear rockets that would intercept and destroy any incoming ballistic missiles. This was a very attractive and superficially plausible idea that promised to make the USA invulnerable. Immediately it was published that prominent scientists with great experience of nuclear weapons pointed out that the whole idea was quite unworkable. It would be easy, for example, for the enemy to equip their missiles with multiple warheads and decoy systems so that the defence would easily be saturated and many of the incoming missiles would reach their targets. A successful defence requires destruction of all the incoming missiles, and this is impossible to guarantee. The whole project was a reckless gamble and a huge expense. It was of course never built, and it could be argued that it forced the Soviets to undertake such expenditure themselves as to make them practically bankrupt, and thus hastened their demise. Nevertheless it is a chastening example of the dire effects of military secrecy.

Much the same can be said for another proposed offensive weapon, the X-ray laser, or Super-Excalibur as it was called. This was intended to focus the intense burst of high-energy radiation from an atomic explosion to form a beam of devastating power. This also was announced with great publicity and was eventually forgotten when it proved completely impracticable.

There is no end to people's credulity; there seems to be a built-in desire to believe in occult forces. Further examples are provided by reports of spoon-bending and psycho-kinesis. Very extensive studies have shown that hidden cards can be identified with a probability just above chance, providing evidence for extra-sensory perception. Further examination does not substantiate such claims. 'In 1987, the National Academy of

Sciences undertook a complete review of all the literature on parapsychology. The report concluded that there was 'no scientific justification from research conducted over a period of 130 years for the existence of parapsychological phenomena' (Park, loc. cit. p. 197). Despite years of failure there are still some people who hope that the next experiment will produce evidence of extra-sensory perception. All this is pathological science.

It should perhaps be pointed out that there is much deliberate fraud, and it might be thought that physicists are the best people to investigate and expose such frauds. This is not so; the best people to do this are professional magicians. We are all familiar with the absolutely amazing things they can do; they produce effects before our eyes that we cannot understand. They know all the tricks whereby these effects are produced, and they are the ones best qualified to expose fraud when it occurs.

Quantum mechanics provides a fertile source of inspiration for a whole range of mystics and new age enthusiasts. They know nothing about quantum mechanics itself, a very powerful and successful theory used everywhere by atomic and nuclear physicists. They have only read some sensational stories about the mysteries of the fuzzy quantum world, the action of the observer in collapsing the wavefunction, the instantaneous transmission of information by non-local forces, the tunnelling of particles through potential barriers and the wave-particle duality of all fundamental particles. This heady mixture of fantasy and erroneous physics fills their fevered imaginations and leads a whole range of nonsense that may indeed look plausible to anyone who knows nothing about physics. It is superfluous to add that quantum mechanics is a method of making objective and exact calculations about the world, and that the nonsense begins when people try to interpret the meaning behind it all. This is still a contentious subject, and it is essential to emphasise that quantum mechanics is a statistical theory that is fully consistent with a completely deterministic world.

A well-known exposure of the decline in scholarly standards in some sections of the academic community is the Sokal hoax. Alan Sokal, a respected physicist, wrote an article called 'Transgressing the Boundaries: Towards a Transformative Hermeneutics of Quantum Gravity' full of outrageous howlers taken from the writings of postmodern commentators on modern science. This article was accepted and published by the journal Social Text. Subsequently Sokal revealed that it was a 'deliberate farrago of deliberate solecisms, howlers and non-sequiturs, stitched together to look good and to flatter the ideological preconceptions of the editors'. It was an elaborate hoax written to expose the appalling decline in standards of truth, reason and objectivity in certain literary and sociological circles. In subsequent writings Sokal and Bricmont provide many more examples of established academics writing complete nonsense about modern science (Bricmont 1997; Sokal and Bricmont 1997; Bricmont 1998, 2003; Sokal 2008).

With such massive evidence of people's ability to believe complete nonsense without paying any attention to the results of careful objective scientific studies, it is no wonder that there are still so many who believe all the nonsense about nuclear power.

8.7. Education

Looking to the future, much depends on the quality of education. The aim must be to ensure that existing and future technological advances are applied for the benefit of society, and that decisions are based on objective facts, uninfluenced by any political or psychological factors.

We are now faced by a wholly new situation. For the whole of human history up to the nineteenth century technology developed as an empirical craft. The skills needed to work wood and metal were passed on from generation to generation, often achieving a very high level. There was no input whatever from the natural philosophers who tried to understand why things behave as they do. Likewise there was no contribution from the universities to the growth of technology, and very little to natural philosophy. Most natural philosophers carried out their work independently of the universities even if, like Newton, they had a university position. For centuries the curriculum of university studies was based on the Greek and Roman classics, and the physics taught was that of Aristotle, even long after he had been completely discredited. In 1868 T.H. Huxley asked a group of very distinguished university men whether it was possible for a student to have taken the highest honours in the university, and yet might never have heard that the earth goes round the sun. Everyone present replied 'yes'. The universities were thus in no position to contribute to the solution of the problems raised by technology.

Now the situation is entirely different. The pace of technological change is so great that new machines are obsolete almost as soon as they appear. Industries grow, flourish and decay like so many mushrooms as their products meet a real need, sell in large numbers and are then superseded. How should education prepare the next generation for this situation?

First of all, there is no need for instruction on how to use the new devices. It is a familiar experience that young children master the use of personal computers and cell phones much more easily than their seniors. The problem with children is the time they spend surfing the internet and what they see there, but such problems are not considered here. Education at the school level can provide guidance, but already many children spend more time with their computers than in the classroom.

Training in the basic engineering, electrical and electronic skills is an essential requirement. This may be provided in technical colleges. The lack of such colleges was the primary reason why Britain gradually lost the technological lead after the great success of the Exhibition of 1851. Germany established institutions like the Charlottenburg Institute of Technology and the USA the Massachusetts Institute of Technology and the lead passed to them. Since then Britain established many Colleges of Technology and of Science and Technology combined, such as Imperial College in London, and the Manchester College of Technology (Ashby 1996). Many new universities were founded in London and the Midland and northern manufacturing cities such as Birmingham, Liverpool, Manchester and Leeds, well equipped with science departments (Ashby 1966).

During the last hundred years the number and size of universities have increased greatly, so that an appreciable fraction of young people aspire to a university education, and even consider it as a right. This has inevitably changed the universities in many ways. In the past there were rather few undergraduates, and the universities subsisted on their fees and endowments. This is no longer sufficient and so the contribution of the State has steadily increased until it now provides the bulk of the funding. Together with the funding comes the danger of political interference. Some politicians are concerned that the proportions of admissions from various social classes are unequal. In particular, the numbers of students admitted to Oxford and Cambridge from public (i.e. private) schools are rather similar to that from State schools, although the numbers educated in these schools are in the ratio of about one to twenty. This is seen as a great injustice and it is suggested that social criteria should be used to redress the balance. The universities point out that they admit strictly on academic merit, and rightly maintain that it would be unjust to act otherwise. The reason for the difference is the excellence of the public schools, and the solution to the problem is to raise the State schools to the same level. This is easier said than done.

Politicians are also under continued pressure to increase the quality of education, and this has resulted in the phenomenon known as 'grade inflation'. Every year the average grades of schoolchildren in the A-levels that control university entrance continue to rise. This is hailed with enthusiasm by the politicians, but everyone knows that this has been achieved by lowering the standards. The result is to give young people an inflated opinion of their abilities, and to make it more difficult for the universities to make a fair selection. The increasing tendency of students to object if they are not chosen, and the possibility of wide and damaging publicity, has also made the task of those responsible for university entrance far more difficult. Another problem is that universities have expanded beyond the numbers of qualified candidates in some subjects. The university is then faced with a difficult decision: either to admit sub-standard candidates to fill the quota or to maintain the standards and face the consequences of reduced numbers. The fact is that already there are far too many students in universities who have no vocation to scholarship but who think that a university degree is a ticket to a lucrative job or a pleasant way of spending a few years while they decide what to do with their lives. This is another way the universities have changed since the nineteenth century.

During the twentieth century the older British universities gradually shook off the legacy of centuries, expanded their science departments and invested more and more heavily in research. The rewards have been prodigious and the new understanding of matter that has been obtained has made possible the new technological advances. One need only mention the discovery in Britain of the electron, of nuclear fission and of the cavity magnetron to show the magnitude of the changes. In the USA the advances have been no less spectacular; the laser, the maser and the transistor provide three examples.

In addition to these scientific triumphs one of the most important contributions made by science to universities is a radical change to the whole process of attaining truth. No longer is truth unquestionably what Aristotle or the Professor says. Evidence must be put forward to support any proposition and it must be defended against attacks. Everyone is equally entitled to join the discussion, and their social standing, qualifications and religion are entirely irrelevant. If it survives this discussion, the proposition may be accepted provisionally until perhaps further considerations are brought forward. This mode of learning may and should now seem obvious, but it was not always so.

Another important contribution of science to the universities is the partial restoration of the universality of learning. During the Middle Ages students moved freely from one university to another to study under the most popular teachers. They spoke the same language and held the same faith and so wherever they went they soon felt at home. Tragically this was shattered at the Reformation: both language and religion became barriers. This situation persisted for several hundred years until science once again unified students and researchers all over the world. Most scientists understand English and are immediately at home among friends and colleagues whether they are in Tokyo or San Francisco, Oslo or Cape Town. In their particular research area, they are familiar with the same publications, carry out research on similar problems with similar methods, and frequently meet each other at international conferences. This invisible college, as it is sometimes called, is geographically wider than in medieval times, and involves far more people, but is restricted to the sciences and subdivided into specialities. In the humanities, which should provide their own unification, the situation is entirely different. Philosophers of different schools will vehemently deny even that they share the same concept of truth.

This universality of science had an important application at the height of the cold war. At that time there was a real danger that some misunderstanding might trigger an all-out nuclear war. This was a general anxiety, but scientists realised that there was a vital contribution that only they could make. Irrespective of political convictions and national alignments, they could agree about certain basic scientific and technical facts such as the dangers of radioactive pollution, the practicability of detecting underground nuclear tests and so on. A meeting between top-level Soviet, American and British scientists was therefore held in the small Canadian village of Pugwash. The costs were underwritten by Mr Cyrus Eaton, a financier. The scientists, many of whom knew each other already, soon established a social rapport. Language was no problem, as many of the Soviet scientists spoke English, and many of the Allied scientists spoke Russian. They discussed only objective scientific facts, and were able to establish a consensus, which they duly transmitted to their Governments on their return. Many such meetings were held over the years, and it is very probable that they injected some scientific realism into the subsequent political decisions.

There are two other respects in which science has made a vital contribution, although they remain far from being understood. The first is to emphasise the importance of numerical evaluation. Many examples in the previous chapters have shown that the truth about many important and contentious matters lies in the numbers. Without the numbers, arguments can go on forever without ever reaching a conclusion; with the numbers, the conclusion is frequently obvious. The second contribution is to show that there comes a time when it is generally recognised that fapp (John Bell's useful abbreviation of the phrase 'for all practical purposes') a particular discussion is settled. Frequently this is the result of increasingly accurate numerical data. After this point, further discussion is just a waste of time. As the previous chapters have shown, endless repetition of the same old arguments, long after they have been refuted a thousand times, continues to bedevil discussions on energy and the environment. Scientists understand this very well, so the major problem is to convey this to students of other disciplines. This is of vital importance because it is predominantly the students of the classics, history and the law that enter public service or Parliament and so are responsible for future policy.

In recent years there has been increasing pressure in several countries to teach science students only the applications of science, not the principles of the science itself. The increase in knowledge is so great, it is argued that time can be saved by concentrating on what is really useful to students. Such a proposal can be made only by someone who lacks even the slightest understanding of the relationship between science and technology. If such a proposal were adopted, the students would doubtless emerge with a comprehensive understanding of how to manipulate the latest gadgets, but they would have no knowledge of the fundamental principles underlying their operation and would be unable to improve or adapt them to changing conditions. Faced with a new gadget, they would be at a loss. The faster the pace of change, the more rapidly would their knowledge be rendered obsolete. It cannot be too strongly emphasised that research into the nature of matter at the most fundamental level and the transmission of this knowledge to students is the primary duty of the university. This is clearly understood by university teachers, and they will not allow this duty to be removed.

In some countries most research is carried out in Research Institutes, and the universities are primarily teaching institutions. This is a perfectly acceptable model, because the university teaching is done by scientists who are often themselves carrying out research in an Institute. If it is argued that all this is still consistent with the students at the undergraduate level being taught only things that are of direct utility, then it may be replied that in that case none of them will be qualified subsequently to undertake research. Furthermore, there is no objection to such courses being given in technical colleges where students are being prepared for practical work in a range of trades, not to undertake research.

In the universities themselves, once it is granted that their primary duty is the teaching of fundamentals and the prosecution of pure research, it may be admitted that the present situation is not always satisfactory, and that possible improvements should be discussed. The main problem is to convey to students of the humanities the need to make a quantitative study of whatever aspects admit of numerical evaluation. The first step is to convince the lecturers of the importance of this, and then it could be illustrated by examples in their lectures. This may not be too difficult for history, but in other subjects it is not so easy. An important contribution could be made by philosophers, who need to leave their preoccupation with logic chopping and sterile absurdities and return to their primary vocation, which is to seek the truth.

The situation in technical colleges offers opportunities for great improvement. It is recognised that these colleges exist to prepare students to apply technical knowledge to the needs of society. For this, they need not only have technical knowledge but an understanding of the wider implications of their work. Thus, in an example quoted by Eric Ashby, those who build a new road in Africa need to realise that this will have implications for the ecology, economy and social structure of the region. People living in what previously were isolated villages can now easily travel to visit each other, to buy and sell produce and to share in a whole range of activities that previously were not possible. Diseases may spread more easily, but medical care is more readily available. A course in road-building technology can then be supplemented by a range of shorter auxiliary courses designed to sensitise students to the likely effects of their activities. The same could be done for the other technical courses.

At the present time there are such powerful commercial forces at work that it is not easy to convince them that there are other considerations apart from their own profit. Even if it is clear that a proposed logging programme or a new dam will devastate the ecology of the region, as well as destroying the lives of the indigenous population it is not easy to get them to modify their plans. This suggests that students at technical college should also have courses to teach them how to fight such companies, backed by the essential legal knowledge.

It is easy to make such proposals, but to implement them in the present political climate in Britain would be practically impossible. The institutions where this might have been done are the old polytechnics. They undertook no research and performed very well the tasks of technical education and training. Then the Government, in response to pressure to increase the access to higher education, decided to transform these polytechnics into universities. This can be done by a stroke of the pen, and the politicians can boast how many new universities they have created. Needless to say, the underlying reality is unchanged. The staff are not equipped for research, although aspirations in that direction have been created. More seriously, there is now a lack of institutions able to carry out the humble but very necessary task of teaching basic technical skills.

The campaign against nuclear power affects the way people think from their earliest years. Children are readily influenced by the mass media, and their teachers may have a political agenda. A survey showed that 58% of 10-year-olds were against nuclear power, but less than half could give a reason. Many children think that nuclear power causes acid rain, that most of the radiation we receive comes from nuclear power stations, and that most children with leukaemia live near nuclear power stations. With this mindset established at an early age, it is not surprising that they continue to oppose nuclear power throughout their lives (Speakers' Corner No. 11. November 1988).

References

Andrew, Christopher M., Mitrokhin, and Vasily (1999) *The Mitrokhin Archive and the Secret History of the KGB* (New York: Basic Books, 2000) (London: Basic Books, 2005), vol. 2 (The KGB and the World. London: Allan Lane).

Ashby, Eric (1966) *Technology and the Academics. An Essay on Universities and the Scientific Revolution* (London: Macmillan).

Bernstein, Jeremy (1993) *Cranks, Quarks and the Cosmos* (Oxford: Oxford University Press).

Bricmont, Jean (1997) Science studies — What's wrong? *Physics World* (December) (Bristol: The Institute of Physics).

Brown, Bob (2004) *Memo for a Saner World* (London: Penguin Books).

Gore, Al (2007) *Earth in the Balance* (London: Earthscan).

Lovelock, J.E. (1979, 1987, 1995, 2003) *Gaia: A New Look at Life on Earth* (London: Heinemann).

Lovelock, J.E. (1983, 1995) *The Ages of Gaia: A Biography of our Living Earth.*

Lovelock, J.E. (2000) *Homage to Gaia* (Oxford: Oxford University Press).

Park, R. (2000) *Voodoo Science* (Oxford: Oxford University Press).

Pears, J. (1998) *An Instance of the Fingerpost* (London: Vintage; Jonathan Cape Ltd).

Sokal, Alan (2008) *Beyond the Hoax* (Oxford: Oxford University Press).

Sokal, Alan, and Bricmont, Jean (1997) The naked postmodernists, *Times Higher*, **10** (October).

Sokal, Alan, and Bricmont, Jean (2003) *Intellectual Imposters* (London: Profile Books).

Stiglitz, Joseph (2002) *Globalisation and its Discontents* (London: Penguin Books).

Chapter 9

The Needs of the Developing Countries

9.1. Introduction

The implications of pollution and climate change are serious for even the most well-developed countries, but they are quite devastating for the poorer ones. It is not acceptable, for both pragmatic and moral reasons, to have a world in which many people enjoy superfluous luxury while others are in the grip of famine, disease and death.

The advances of science and technology have provided the means to tackle this situation; what is lacking is the will to take the necessary action. Even with the necessary will, there are formidable obstacles to be overcome, in particular with those connected with the economic, political and psychological situations in the poorer countries. These interact with each other and differ from one country to another, but there are many features common to all. These problems are so complex that it is not possible to do more here than to sketch a few ideas.

Many of the problems of the developing countries have already been discussed in the previous chapters. Thus one of the most serious problems is the shortage of food. Some developing countries are forced by their mounting debts to grow cash crops for export, instead of food for their own people. The worldwide neglect of nuclear power has exacerbated the growing shortage of oil, thus greatly increasing its price, and this bears most heavily on the poorer countries. This is now being made worse by growing crops to produce biofuels, and as discussed in Section 3.9 this is raising the price of food still further.

153

The tribal wars endemic in many countries, especially in Africa, have led to their spending vast sums on importing weapons instead of building hospitals and schools and investing in more efficient agricultural machinery and fertilisers. The loss of life and the damage caused by these wars worsen their situation. The breakdown of tribal authority has lowered moral standards, leading to the spread of AIDS and other diseases.

9.2. Economic Problems

The countries most in need are the least able to afford the vast sums necessary to provide the infrastructure that is essential for their development. They lack the technical expertise and the trained manpower to build their own power stations, and so have to import them from abroad. The same applies to all the other large-scale industrial equipment that will enable them to stand on their own feet and not be dependent on aid and imports. Many of the poorer countries have huge debts and so have to devote much of their agriculture to cash crops to repay even the interest on the debt instead of growing the food they desperately need.

Generous aid from the richer countries is needed, but it must always be with the object of eventually securing the economic independence of the recipients, and discouraging a complacent begging-bowl mentality. Arrangements whereby foreign companies are encouraged to build power stations and factories as a long-term investment with the eventual aim of transferring ownership to the recipient country may provide one way of getting the economy started. The raises delicate political problems, as such arrangements are often seen as economic imperialism. A condition of their success is that such investments must be protected against unjust taxation and confiscation. The willing co-operation of the recipient country is essential for success.

In the case of highly sensitive industries, such as nuclear power stations, it is necessary to ensure that they are not used to manufacture weapons. In countries with unstable political regimes, this might be done under the auspices of the International Atomic Energy Agency, supported by the United Nations.

In some countries African dictators have illegally amassed huge personal fortunes that are now in Swiss banks. It has been estimated that such sums, equivalent to many billions of pounds, are similar to the total debts of all African countries. Means must be found to release these sums and make them available for development.

The economies of many poorer countries have been severely harmed by programmes of nationalisation and by the confiscation of farms owned by immigrants or their descendants. These farms are given to politically approved people who lack the knowledge to run them, so production falls and everyone suffers.

The economic problems of many countries are further exacerbated by the rapid growth of population due to improved medical care. Parents welcome large families as they help to safeguard their future. This is now offset by the growing scourge of AIDS which leads to growing numbers of orphaned children which places an increased burden on already overstrained medical resources.

One of the greatest needs of the developing countries is for more energy. As already mentioned, this is essential at both the industrial and the domestic level, and has a strong correlation to health and general well-being and quality of life. The population of the larger cities is rapidly increasing, and many of the largest cities in the world are in the poorer countries: Mexico City with 25 million, Calcutta and Cairo with 20 million and countless cities with a million or more. The demand for energy in developing countries is likely to grow by 150% by 2030 and globally to more than double by 2050. Only nuclear power is able to provide this energy without serious pollution. I recall attending a conference in Calcutta some years ago and the pollution caused by about six million wood and dung fires was such that about one-third of the participants experienced bronchial problems.

The need for energy can be met by nuclear power, but then there are to problems of paying for it and preventing the diversion of fissile material for illegal purposes. Most of the developing countries are already heavily in debt, and have urgent problems that they have to tackle with their limited means. One possible way to solve this problem is to encourage companies, under the auspices of the United Nations, to build nuclear power stations at their own expense and then sell the electricity produced. It may be desirable to designate the land on which the power stations are built as United Nations land, and keep it well guarded to prevent sabotage or theft of fissile material. Such reactors could also produce radioisotopes for medical purposes.

Most nuclear power reactors are 1000 MW; this is excellent for large cities but not for smaller ones, especially in countries without a well-developed grid. These can be served by smaller reactors of 50 to 500 MW. They are inherently very safe and easy to operate, and also very secure. They have long fuel load lifetimes that may be sufficient to last throughout their life, so that refuelling is not necessary. Such reactors have been proposed by the Global Nuclear Energy Partnership (GNEP), an international consortium of the leading nuclear nations, namely the US, France, China, Russia and Japan. Several countries have extensive reactor research and development programmes; for example France spends about $500M per year and Japan over $2000M per year. Shamefully, the UK has practically ceased such work, and so is present at GNEP meetings only as an observer (Nuclear Issues, September 2007).

While there is a general agreement that in order to avoid serious climate change the emission of carbon dioxide must be controlled, there is a disagreement about the reductions to be undertaken by different countries. The poorer countries argue that the industrialised nations have developed their industries taking no account of the pollution they have caused, so that it is unjust to retard their own development by imposing reductions on them. The developed countries find it difficult to accept this because of the extra cost that this would impose on their industries, which would then be economically disadvantaged. The arguments centre around proposed reductions by 10% or 20% in a few decades, whereas the reality is that every country must reduce their emissions practically to zero if serious consequences are to be avoided.

In all countries there are rich and poor people, but it is particularly scandalous that this is also true for many developing countries. In the big cities there are areas with vast opulent mansions where the richest people live in luxury, waited on by countless servants. Not far away, and also in the surrounding towns and villages, there are millions living in abject poverty in crowded shanty towns with no access to water, electricity or basic health care. They are well aware from discarded magazines and television of the high standard of living of the rich and are filled with a burning sense of injustice. This is an unjust and potentially explosive situation that calls for urgent action.

9.3. Political Problems

In many cases poorer countries have recklessly squandered their natural resources and manpower by indulging in murderous wars. It is the first step on the way to development to stop these wars. In many continents tribal wars have been endemic since the dawn of history, and they are made far worse by modern weapons. In many cases the infrastructure, the roads and communications, the buildings and transport systems that have been laboriously built over the decades are wantonly destroyed.

The situation in many African countries is sometimes blamed on colonialism. Undoubtedly many countries were cruelly exploited in the past, but it must be added that much was also done to support development. In any case, many years have now passed, and other countries, particularly in South-East Asia, which also suffered from colonialism, now have booming economies and great industrial strength.

The educational systems and policies in many countries are not conducive to development. There is a great demand for education as it is rightly seen as the way to a better future. However, education is often seen as following the European model with its emphasis on the liberal arts. No one would want to dispute the value of the liberal arts, but what is also needed is technically trained manpower to build and maintain the industrial base and the infrastructure that goes with it. The young men and women predominantly aspire to be lawyers, politicians and economists, and not medical doctors, scientists and technologists.

The educational situation in some countries is made worse by what is called preferential access. This means that in a country with several racial groups preference is given to members of the indigenous population rather than to those of immigrant groups that may have been in the country for hundreds of years. The result is that many highly qualified young men and women, seeing no future in their own country, are forced to emigrate to countries that offer more chance of employment, thus depriving their country of the skills that it so desperately needs. In one such country, the only professions in which applicants are judged by objective criteria are surgeons and airplane pilots. Not even the most fervid politician wants to be operated on by someone who does not know what he is doing or to fly in a plane with an unqualified pilot. Inevitably, the policy of preferential access leads to a gradual lowering of the standards of universities and other professional occupations.

9.4. Psychological Problems

An essential pre-condition of development is an attitude of mind that recognises the value of work as a preparation for the future. Every population has people with a spectrum of attitudes from lazy indolence to hard-working activity, and much depends on the relative numbers in each category. In countries where most people are honest and hard-working thrive, whereas others remain poor, it is necessary to face the unpopular truth that the situation in many of the poorer countries is due to a general lack of a strong work and save ethic. Each day is enjoyed for its own sake, with no thought for the future. The result is that no one prepares for the future. Machines are not properly maintained and when they break down there are no spare parts available or people who know how to get them going again.

This is illustrated by the present situation in South Africa. About ten years ago it was proposed to build some nuclear power stations, but this was prevented by the opposition of the Greens, who said that all the power needed could be obtained by building windmills in the Karoo desert. The growing demand for energy has now put an unsustainable burden on the ageing power stations, and as a result frequent unannounced power cuts are a daily experience. Suddenly the lights go out and everything powered by electricity stops. The mines have had to close down, leading to serious loss of revenue. Supermarkets are paralysed, the refrigerators cut out and fish and meat have to be thrown away. The traffic lights fail, leading to traffic jams and increased risk of accidents. A power failure caused the cable car on Table Mountain to jump off the rails, stranding many passengers for several hours. Many businesses have installed emergency power generators, and the cost of this is passed on to the consumers. All this will happen before long in many well-developed countries unless their energy policies are radically changed.

Chapter 10

Moral Problems and Responses

10.1. Introduction

Since scientific advances can be applied in many different ways, some good and some evil, technology raises many moral problems. The development of new means of generating energy is no exception. A new technology may displace a well-established one; this benefits the whole community and provides new opportunities for employment, but will inevitably destroy the jobs and lifestyle of those working in the older industry. This is no argument against the new developments, but it requires concerted action to alleviate the distress of the unemployed workers, and ensure that whenever possible they are re-trained for other work.

Many new technological advances can be used both for good and evil purposes. Thus nuclear power stations can be used to produce electricity and also to make fissile material for bombs. This is no argument against a new technology, for otherwise it would be immoral to make knives. It is different if the technology has only an evil purpose and there is continuing argument about whether this is the case for nuclear weapons.

It is much the same for all methods of generating energy, for as described in Chapter 5 there is no perfectly safe way to do this. Whatever method we use will involve the likelihood of people being killed and injured. Certainly we are obliged to reduce these as far as we can, but it is not possible to eliminate them entirely. It is not only the workers who are concerned as there are usually deleterious effects on the whole of society, particularly in the case of fossil fuel power stations.

The responsibility of choosing the optimum choice of energy sources lies first with the Government and all those involved in the design, construction, running and maintenance of these energy sources. Moral problems can arise even for a scientist asked to work on a fundamental problem. He may know that it is the intention of his sponsor to apply it for a purpose that he considers immoral. He may then consider that it is his moral duty to refuse to work on such a problem. I knew one such case where the scientist refused, and another scientist undertook the work. As it happened, the result proved useless for the hoped-for immoral applications but had many useful and legitimate applications. It was one more illustration of the general experience that it is impossible to foresee the results of any piece of research.

Intractable moral problems arise when a new technological advance renders an existing industry obsolete; such cases are considered in Section 10.2. The moral questions require the guidance of Church leaders, and their responses to the energy problem are considered in Section 10.3. Some misleading statements about energy questions are listed in Section 10.4 and conclusions are drawn in Section 10.5.

10.2. Industrial Changes

Many of the most difficult problems concerned with the development of technology arise when a well-established industry is suddenly rendered obsolete by a new invention. There are many examples of this, such as the numerous improvements in machinery for spinning cotton, the replacement of horse-drawn stage coaches by railways and automobiles, and canal transport by the railways, and the replacement of large passenger liners by huge jet planes. Other dramatic changes, such as the almost continuous development of new technology for war, the introduction of gunpowder and the atomic bomb, have different effects as they are organised and paid for by the State.

In the nineteenth century there was a flourishing trade in ice between the USA and India. In the winter, blocks of ice were cut from the frozen lakes in New England, packed in wood shavings and taken by sailing ship to Calcutta. Surprisingly enough, sufficient ice survived the journey to be available to make ice cream during the hot summer. This useful commercial activity was killed stone dead as soon as refrigerators were invented. Another example is provided by indigo, a dye that was highly prized from ancient times. In India, two million acres were devoted to growing the plant from which it was extracted. The chemist von Bayer succeeded in 1883 after twenty years to synthesize the dye, but the process was too complicated for industrial development. Then one day in 1896 a careless worker stirred a mixture of naphthalene and sulphuric acid with a thermometer; the thermometer broke and the reaction took a different course, producing the vital ingredient for the synthesis of indigo. The mercury acted as a catalyst for a hitherto unknown reaction. Indigo could then be synthesized, and the Indian indigo industry was destroyed (Gratzer 2002, p. 46).

The rendering obsolete of an established industry has devastating effects on the lives of the workers. Their skills, honed over years of apprenticeship and practice, are suddenly made useless. The factory or mine where they work is forced to close down by inexorable economic pressure. They may lose their homes as they are no longer able to pay the rent. There may no longer be any jobs for them in the neighbourhood. Their situation is indeed desperate.

The changes may be foreseeable, and may take place gradually. It is then possible for a sympathetic management to explain what is happening, provide help for the workers to move to another district where jobs are available or even re-organise the factory to produce new goods that are saleable. Some workers may react violently and smash up the new machinery that is threatening to destroy their jobs, as the Luddites did in the north of England.

Another type of difficulty arises when the Government or the industry itself insists, for good reasons, on new safety precautions. These inevitably increase the price of the product and may make it uncompetitive compared with similar industries in other countries, which are not subject to the same regulations. This situation may be worsened if the workers in the other countries are willing to work for longer hours for lower wages. It is then impossible for the first country, which may have developed the industry in the first place, to survive the competition. In this way the production of cotton cloth in England was destroyed by the competition from Asiatic countries such as India. The same thing happened later on to the British shipbuilding industry, at one time the greatest in the world.

The steady improvement in manufacturing methods and the development of new products is ultimately responsible for widespread benefits and the raising of living standards. However it often has devastating effects on the lives of the workers concerned. In the nineteenth century many trade unions were formed and helped to negotiate better working conditions and remuneration. In many cases they organised strikes that caused loss of production and sometimes destroyed the whole industry. A contributing cause in many cases was complacency among the management and a reluctance to introduce new techniques.

As an example, the Japanese shipbuilding industry was completely destroyed by bombing during the war, and subsequently they were forbidden to build ships. They therefore assigned many young engineers to study shipbuilding and to learn from techniques used in other countries. They developed new and faster techniques, so that when finally they were permitted to build ships they could do so rapidly and reliably, using the latest methods. The British shipbuilding industry, wedded to older techniques and crippled by strikes organised by Communist trade unions, could not compete and one after another the great shipbuilding yards were forced to close down.

So competitive is modern manufacturing that the rights of the workers are often ignored, and they are treated simply as units of production. This is an affront to their dignity as human beings, and over the years the Church has insisted that their rights are respected. A whole series of Papal Encyclicals, Rerum Novarum in 1891, Quadragesimo Anno in 1931 and Tertio Millennio Adveniente in 2000, together with many addresses on various occasions, have described the rights of workers and how they may be implemented.

10.3. The Contribution of the Churches

It is a characteristic of human beings that we tend to take effective action only when it is too late. The problem of continuing to supply industrial countries with enough electrical energy to power their various activities has been evident for years and many scientists have issued statements warning everyone of the crisis that will soon arise as the oil and then coal are exhausted. Details of the technical aspects of this problem have been given in Chapter 3. Here we are concerned with the moral problems. Many Churches have rightly considered that this is a moral as well as a technical problem, and have organised conferences and issued statements. In this section these official reactions of various Church authorities are summarised and reviewed. In addition to the official statements many individuals have formed groups and issued statements of their own. These have often influenced the official Church actions.

The role of the Churches in the energy debate has been considered in an article by Kenneth Fernando of Sri Lanka in a booklet 'Energy for my neighbour: Perspectives from Asia' edited by Janos Pasztor and published by the World Council of Churches in 1981. He begins by recognising the reality of the energy crisis and the limited influence of the Churches. Nevertheless he maintains that the Churches must use whatever influence they have not only by discussing the ethical issues but by actions that use its source of power, 'the power of the people, the power of suffering and the power of God at work in the world'.

'The first task of the Churches is to work for a better understanding — a new understanding — of man's relationship with nature. The book of Genesis teaches that man has been endowed with "dominion" over all the earth.' Emphasis on this aspect has 'led to an unbalanced doctrine of man's relationship with nature'. Thus 'an aberration in the Western Christian ideology has led to a theology that justifies a rapacious attitude towards the gifts of nature'. 'Traditional Asian thought is very different. Man is a part of nature and therefore cannot seek to master nature. He must live with it, in harmony with it. He is entitled to use nature, but not to abuse it. He is entitled to develop it, but not to exploit it. This idea of living in harmony with nature is by no means alien to the Judeo-Christian tradition.'

'Another task which the Churches must perform is to fulfil their prophetic role. It is the duty of the Churches to speak out like the prophets of old, whether they are heeded or not, whether it makes them popular or not.'

He concludes his article with some practical suggestions. We must rethink our lifestyle so as to use less energy and no more than our fair share of the earth's resources. 'The Churches cannot remain neutral in the face of the energy crisis. They must take sides with the poor, since they are the worst affected .... Protests against injustice generally are not likely to be of much avail. It is necessary to focus attention on one or two specific issues and to mobilise support for them.' As examples of this he cites Mahatma Gandhi's advocacy of 'spinning and weaving in India, to draw attention to the injustices of a dependent Indian economy', and 'the issue of infant foods to expose the fact that multi-national corporations create artificial needs in the third World in order to satisfy them'. Similar actions must be taken in relation to the energy crisis.

It is not easy to comment on technical subjects in non-technical language. The choice between energy sources is not a subjective matter like assessing the merits of two pieces of music. The capacities, costs, and safety of the various methods of energy generation must be expressed numerically, as this forms the only basis for a rational discussion. Naturally there are uncertainties associated with these figures, but these can themselves be treated by familiar statistical techniques. There are some rather subjective aspects, especially when it comes to assessing the effects on the environment, but basically it is a quantitative discussion that must be conducted in quantitative terms. If this is not done, it becomes just a matter of emotion and rhetoric. It is fatally easy to put together true statements so as to convey a totally false impression, and this is the technique used by the propagandist who wishes to provide support for policies already decided on quite different grounds.

The remainder of this section is devoted to accounts of a representative sample of Church statements on nuclear power. In some cases it is necessary to make critical comments, so it is appropriate to emphasise that I fully accept that the people responsible for them are dedicated individuals whose only aim is to serve the community. Unfortunately the highest motives do not guarantee freedom from error. No good purpose would be served by covering this up, but no criticism of the authors is implied by critical analyses of their statements.

1. The Church of England

Nuclear Crisis. Edited by Hugh Montefiore and David Gosling (Prism Press, 1977), p. 165.

This is a serious and substantial attempt to tackle a particular problem, namely whether fast breeder reactors should be developed in Britain. According to the Foreword, it is 'part of a world-wide effort by the Churches to respond to the challenge of new technology.' The initiative came ultimately from the World Council of Churches, whose General Assembly requested the British Council of Churches to convene public hearings. They in turn asked Bishop Montefiore and Dr. Gosling (a nuclear physicist) to arrange the hearings. The book contains an edited account of the submissions to an expert panel and the cross-examination of expert witnesses at the Public Hearings on the Projected Commercial Fast Reactor CFR-1 held in London in December 1976.

The book contains a considerable amount of detailed information, essentially all of which is available elsewhere. There is much expression of contrasting opinions, but rather little attempt to decide what is right and what is wrong. Bishop Montefiore repeatedly emphasises that 'our aim on this panel is not to pass a verdict but to expose arguments and assumptions.' On the dust jacket it is written that 'the hope is that on reading the cut and thrust of debate the public will be able to make up its own mind on whether the nation should commit itself to a long term nuclear programme.' This is wishful thinking. How many of 'the public' are likely to read this book, and of these how many are able to make an informed judgement? If the distinguished and expert members of the Panel cannot make such a judgement, what hope is there that members of the public will be able to do so? In saying this it is not implied that the Panel should have been able to give a direct yes or no to the question of whether to build CFR-1. This depends on economic and political questions that are changing and possibly difficult to decide. But they could have clarified the essentially Christian contribution to the debate, in particular with reference to considerations of safety and securing our future energy supply. Although it was perhaps outside the terms of the enquiry, reference could usefully have been made to the impact of our decisions on world energy supply, and in particular the effect on people in the poorer countries. They had the material to make a statement of real value, but failed to do so. Decisions have to be taken now, and on them depend the lives of millions, whether they starve, whether they die. This is the way the world is. Are the Churches willing and able to participate in taking these decisions?

Nuclear Choice: A Christian Contribution to the Debate on the fast Breeder Reactor. An Occasional Paper of the Board for Social Responsibility (CIO Publishing, London, 1977), p. 8.

The Board for Social Responsibility considered that the most useful contribution that it could make to the decision whether or not to build a commercial fast reactor was to issue a short pamphlet which would throw light on the nature of the choices we face. It emphasised that the pamphlet must not be thought of as stating the view of the Church of England; rather it is a 'reflection on these questions in the light of a Christian concern for the right use of the earth's resources, and a proper sense of our responsibility to those who come after us'.

The pamphlet uses the material collected at the public hearings on the question, and published in the book Nuclear Crisis. It is essentially a brief summary of the main questions, with no attempt to distinguish between truth and rubbish. It abounds in statements of the type: 'some say this and some say that'. Some of the statements are simply false, for example: 'Wave power research is being fully supported, and mass production could begin in 1985.' In fact, of course, wave power is not currently capable of meeting our energy needs. The final section on 'The Churches' contains a choice selection of misleading platitudes such as 'the experts disagree', 'the public must be informed' and a demand for more time for issues to be clarified before major decisions are made.

If all that the Churches can do is list arguments in this way, with no attempt to weigh them and come to a conclusion, it is no wonder they are ignored. Far from being responsible, it is the abnegation of responsibility.

It is greatly to the credit of the Church of England that it took the issue of nuclear power very seriously, and devoted much time and trouble to collect information. But this is only a preliminary to a careful and critical analysis of the information, followed by a decision about what action should be taken. The distinguished Panel that conducted the Hearings unfortunately did not go beyond the first of these three steps.

Quite recently, Bishop Hugh Montefiore and Bishop John Oliver have made statements strongly supportive of nuclear power. Impressed by the mounting evidence for global warming Bishop Montefiore concluded that 'the solution is to make more use of nuclear energy', and resigned from his position as a trustee of Friends of the Earth, which he had supported for many years. He added: 'It is crucial if the world is to be saved from future catastrophe that non-global warming sources of energy should be increasingly available after 2010'. He went on to summarise and refute the objections often made against nuclear power and concluded: 'The advantages far outweigh any objections, and I can see no practical way of meeting the world's needs without nuclear energy.

The predictions of the world's scientists are dire, and the consequences for the planet are catastrophic. That is why I believe that we must now consider nuclear energy'. This statement is particularly impressive because for many years Bishop Montefiore has been actively concerned with questions of the morality of nuclear power. Bishop Oliver, in his contribution to the Belmont Abbey Conference on environmental questions in 2004, said that: 'I am personally absolutely convinced that we have to have another generation of nuclear power stations.' At the present time, he went on, 'the worst risk is climate change, the second worst risk, I think, is the interruption of our supplies of natural gas. The risk associated with nuclear, I think, is less than those; that is my personal opinion. I am sorry, personally, that the Government White Paper did not say we have to go straight forwardly for the new generation of nuclear power; I think we have to'.

2. The Methodist Church

1. *Shaping Tomorrow.* Published by the Methodist Church Home Mission Division (1981). Edited by Edgar Boyes, p. 72.

This is a consideration of several modern technologies including electronics, nuclear power and medicine, together with a discussion of the moral issues in the light of Christian principles. Chapter 3 is devoted to nuclear power and contains discussion of the energy problem, nuclear reactions, God's purpose in nature, energy resources, risks and social implications. The treatment throughout is well-informed, accurate, illustrated with well-chosen statistics, balanced and comprehensive. The conclusions are worth recording in their entirety:

1. Nuclear energy is an integral part of nature, just as much God's creation as sunshine and rain.
2. It does offer mankind a new energy source which is very large, convenient and not very costly.
3. Around the world the most important energy sources, oil in the rich world and wood in the poor, are becoming scarce, so that we cannot afford to set aside any energy technology with a large potential which is cost effective, provided it is reasonably safe.
4. There are risks associated with the use of nuclear power, as with everything else, but these have been very carefully evaluated are

not very big and are not at all out of scale compared with risks of other energy sources and other ordinary hazards.

This document is a gem that deserves the widest circulation. It is the only reliable and readily-available discussion of nuclear power in the light of the Christian faith.

It is not difficult to understand why this report achieves such a high standard. It was based on two and a half years of work by a group of some sixty scientists, technologists and engineers including many from UKAEA, Harwell. This provided the massive scientific and technical expertise essential for such studies.

It should however be noted that this report is not an official statement of Methodist belief, and it indeed aroused some controversy. A friend of mine, a retired professor of nuclear physics then living in Australia, tried to get a copy of Shaping Tomorrow and wrote to me that 'a friendly and helpful minister of the 'Uniting Church' (formerly Methodist) told me that he thought I would have difficulty because the views propounded were contrary to the policy of the Church here. He was quite right, but I later obtained copies of the booklet from London. I find it very sound and sensible on three questions: genetic manipulation, nuclear energy and microchip technology.' My friend took a copy with him on a visit to New Zealand and 'was amazed and pleased to see with what pleasure and astonishment my intelligent friends read it. They were mostly Methodist but they were taught quite different views in New Zealand'.

It is not surprising that Shaping Tomorrow attracted such opposition within the Methodist Church because it goes against so much anti-nuclear propaganda that is widely accepted. The writers of the Report certainly performed a signal service not only to Methodists but to all Christians.

If the Methodist Church had only produced Shaping Tomorrow, and had devoted its further energies in this area to disseminating, developing and applying its conclusions, it would have deserved the highest praise. However, the Alliance of Radical Methodists were strongly opposed to Shaping Tomorrow, and produced a further report dealing with the same issues from an alternative point of view. This Report was also published by the Home Mission Division of the Methodist Church, although it felt bound to dissociate itself from some of its content and method. This publication will now be discussed.

2. *Future Conditional: Science, Technology and Society — A critical Christian View*. Published by the Home Mission Division of the Methodist Church (1983). Edited by B. Jenner, p. 96.

This report is a wide-ranging analysis of the relations between science, technology and society, with particular reference to the situation in Britain. It is written from a radical socialist standpoint, and is thus opposed to the whole capitalistic basis of our society. After a general introduction to the deficiencies of our society, successive chapters are devoted to medicine and health, food and farming, energy, arms production, electronics technology, work, wealth and power, faith, science and the human future, and finally the nature of science. Here we discuss only the chapter on energy. Most of it is devoted to mainly qualitative discussions of the various energy sources and their political and environmental aspects. There is much optimistic enthusiasm for the renewable energy sources, and a long discussion of the disadvantages of nuclear power. Each section is headed by a symbol: that for the section on nuclear power is the mushroom cloud of the atomic bomb. Detailed quantitative comparisons of the costs and safety of the various energy sources are conspicuous by their absence, and yet it is only possible to base balanced decisions on this information. There are many references to the hazards of radioactivity, but no quantitative comparison of dose rates from various sources.

The authors of this document are very dissatisfied with the present policies but do not support their alternatives with sufficient factual argument to carry conviction. If one disagrees, as they do, with Shaping Tomorrow, it is essential to try to show where it is incorrect in its detailed scientific analysis. This is where the battle must be fought. One must try to show, for example, that solar or wind power is cheaper and safer than nuclear power. This is not attempted; instead, we are given a mixture of wishful thinking and socialist rhetoric. The same applies to their advocacy of the low energy scenarios, which are not examined in sufficient detail to show their disagreeable consequences. The desire to revolutionise our whole society seems to be of more concern to the authors than the need to feed the poor now. It is not realistic to rely on possible future developments of unpromising energy sources; we must use the means available to us now.

Even judged as a presentation of the socialist case, this is a remarkably poor document. The reason is clear when we look at the list of authors and their qualifications. There is not one professional physicist, and the chapter on energy was written by an engineer who works on solar and wind energy. With the best will in the world, such a group is not equipped to tackle such difficult scientific and technical problems.

It is almost unbelievable that a report of such quality should be published by the same organisation as that responsible for Shaping Tomorrow. To their credit, the publishers were unhappy about it, as they say in the Preface that 'the Division feels bound to dissociate itself from some of the content and method', although it does not say which content and method. It is unhappily inevitable that Future Conditional will undo much of the good done by Shaping Tomorrow. Indeed, it is very likely to have more impact on the young and impressionable by the way it is presented. Was the Home Mission Division of the Methodist Church incapable of distinguishing between excellence and incompetence, between objective scientific analysis and socialist propaganda?

3. The Catholic Church

Several statements have been made by Catholic Episcopal conferences and are published in the volume 'European Churches and the Energy Issue'.

In addition, the Pontifical Academy of Sciences organised a Study Week on 'Mankind and Energy — Needs, Resources, Hopes', and the bishops of the United States issued a detailed statement on the energy crisis.

1. *Energy and Mankind-Needs, Resources, Hopes.* Proceedings of a Study Week organised by the Pontifical Academy of Sciences and held in the Vatican City from 10–15 November, 1980. Edited by Andre Blanc-Lapierre, Pontificiae Academiae Scientiarum Scripta Varia (46), 1981.

At this meeting about thirty leading authorities on various aspects of energy production presented papers and discussed the moral implications. The result is a very detailed, authoritative and substantial (719 pages) study of the whole energy problem, a mine of information on practically every aspect of the production, distribution and use of the various energy sources. Particular attention is paid to the safety aspects.

In the Conclusion to this meeting it was emphasised that 'at the present stage of world development it is not possible without additional energy availability to cope with the population growth, increasing demand for food, and with the problem of unemployment: furthermore, a lack of energy can indeed menace world peace'.

The present world economic situation is then described, with particular emphasis on the increasing dependence on oil, and the economic consequences of the steep rise in oil prices in the 1970s. This threatens the stability of the existing economies, and puts 'the non-industrialised countries in an extraordinarily vulnerable position'.

Urgent action is needed. 'We have no time to waste. Energy policies are urgently needed, involving concerted action by the responsible bodies, and this requires the support of public opinion and energy users. Unfortunately, even in the industrialised countries, the public consciousness of the problem is lacking.' The industrialised and oil-exporting countries must 'help the poorest countries to develop their own energy resources'.

'Only coal and nuclear power, together with a strong energy conservation policy and continued gas and oil exploitation and exploration, can allow us to effectively meet the additional needs for the next two decades. It is emphasised that the industrialised countries must reduce their oil consumption and leave it essentially for specific end uses (transportation, petrochemistry etc), and for the basic needs of the developing countries.'

'No energy source should be neglected if we wish to resolve the energy crisis. A strong research effort must be made to develop renewable energy sources which, among others things, can encourage decentralisation of human settlements, thus reducing the disturbance of the excessive urbanisation process that has occurred and is still occurring in the world.'

'A mix of energy resources and technologies is essential to reduce the vulnerability of the socio-economic system, and to retain the necessary flexibility, thus being prepared to cope with unforeseen events like sudden changes in source availability'.

'Particular attention was paid to the possible consequences of an increase in the carbon dioxide content of the atmosphere. As carbon dioxide effects on climate are not yet completely understood, continuing extensive researches on climate, on photosynthesis and on the photochemistry of carbon dioxide fixation should be pursued so that possible detrimental affects of carbon dioxide may be detected early enough to take effective actions'.

'As regards the use of nuclear energy, some concern has been voiced as to the possible links between nuclear energy and the proliferation of nuclear weapons. In this field, however, it is recognised that, once a certain level of knowledge and technical expertise has been acquired, a country's development of nuclear weapons is primarily determined by political considerations. Thus, with adequate precautions, there is no reason to bar the development of nuclear energy for civil use'.

The safety and health of those engaged in the production of energy should be taken seriously, as well as the safeguarding of the environment.

It should be possible to ensure adequate energy before or about the turn of the century, provided action is taken now. 'Intensive research and development action can make it possible to satisfy the long term energy needs of mankind using the vast reserves of coal, non-conventional oil and gas, uranium in breeders and renewable energies.'

'The recent growth rate of energy consumption in the industrialised countries cannot continue indefinitely. To resolve the present crisis is not sufficient. It is also necessary for these countries to evolve new less energy consuming ways of life, which will promote new patterns of development.'

'All nations have become interdependent, not only insofar as growth rates are concerned, but also with respect to raw materials, agricultural products, technologies, and the knowledge necessary for development. This interdependence and the problems that it poses emphasise the necessity of new kinds of cooperation between nations.'

Scientists are responsible for evaluating the data, and political leaders must take decisions adapted to the needs of both current and future generations, supported by engineers, sociologists and churchmen and indeed all who can influence the future. 'Cooperation between these groups is highly desirable, at the national and even international level, especially insofar as it brings out the human and hence ethical dimension of energy issues.'

The conclusions of this study week of the Pontifical Academy have an authority beyond that of their authors. In this case they were given additional authority by being presented as the contribution of the Holy See to the International Conference on Nuclear Power Experiences organised by the International Atomic Energy Agency and held in Vienna from 13–17, September 1982 (Document IAEA–CN–42/449). The conclusions of the study week were endorsed by Mgr Mario Peressin, the Permanent representative of the Holy See to the IAEA, in his address to the conference. The central feature of his speech was the firm statement that 'my Delegation believes that all possible efforts should be made to extend to all countries, especially the developing ones, the benefits contained in the peaceful uses of nuclear energy'.

It is however notable that the work of the Pontifical Academy has been almost entirely ignored.* It is admirably suited to be the source book of information on the energy problem that further studies could use as a basis, but it has not been used in this way. Thus a conference on 'The Christian Dimensions of Energy Problems organised by the Catholic Union and the Commission for International Justice and Peace and held in Brunel university in April 1982 made no reference to the work of the Pontifical Academy. Thus a great opportunity to propagate sound views on energy problems was lost.

The growing public awareness of the hazards of pollution and climate change has prompted many additional church statements. Notable among them is 'The Call of Creation' issued by the Department of International Affairs of the Catholic Bishops' Conference of England and Wales in 2002. This is an eloquent and urgent reminder of our responsibility to care for the environment and not to squander the resources of the earth, supported by quotations from the Scriptures. It reminds us that we are already using more than our fair share of the resources of the earth, and our consequent obligation to reassess our lifestyles and to moderate our consumption. Its impact is however lessened by the lack of specific examples of the practical measures that should be taken by the individual and the State. In particular it makes no mention of the urgent and specific recommendations it made by the Pontifical Academy.

4. The World Council of Churches (WCC)

The World Council of Churches has devoted serious attention to the problems of nuclear power, and a chronology of the Church and Society

---

*An exception is the article Nuclear power: Rome speaks. *The Clergy Review* (February 1983), vol. LXXVII, p. 49.

Meetings and Publications from 1976 to 1982 is given in the July 1983 issue of Anticipations. Among these are the following documents:

(a) Facing Up to Nuclear Power. Edited by John Francis and Paul Abrecht (1976).
(b) The Churches and the Nuclear Debate. Anticipation, No. 24 (November 1977).
(c) Energy for My Neighbour (1978).
(d) Equations for the Future. Anticipation, No. 26 (June 1979).
(e) Faith and Science in an Unjust World. MIT Conference (1980).
(f) Energy for My Neighbour. Anticipation, No. 28 (December 1980).
(g) Energy for My Neighbour: Perspectives from Asia. (ed.) J. Pasztor (1981).
(h) Alternative Energy Paths; Utopia or Reality? Anticipation, No. 29 (November 1982).
(I) Hope for the Future. Anticipation, No. 30 (July1983).

This list shows that over the years the World Council of Churches has devoted much care and attention to the problems of the energy crisis in general and nuclear power in particular. Many meetings have been held, many experts consulted and many books and articles published. Taken together, this work constitutes the most extensive of all the Church contributions to the nuclear power debate. Particularly valuable and commendable is the constant concern for the poor, ever the mark of the Christian. It is not practicable to summarise all this work here; instead some highlights and failings will be discussed.

In the book 'Facing up to Nuclear Power' Alvin Weinberg identifies the vital issue, namely that scientists and technologists can build nuclear power plants, and can say that if care is taken the likelihood of a serious accident is extremely small, but 'society must then make the choice, and this is a choice which we nuclear people cannot dictate. We can only participate in making it. Is mankind prepared to exert the eternal vigilance needed to ensure proper and safe operation of its nuclear energy system? This admittedly is a significant commitment that we ask of society. What we offer in return, an all but infinite source of relatively cheap and clean energy, seems to me to be well worth the price'.

A general failing is the inability to recognise the real experts and accept their views and to prefer instead the results of a democratic vote by people unfamiliar with the detailed technical knowledge that is necessary to make possible a worthwhile contribution. An example of this is contained in the paper on 'The Churches and the Nuclear Debate'. This contained a magisterial survey by Hans Bethe, a very distinguished nuclear physicist, who presented a survey of the nuclear debate: 'It must be realised that there is a continuing energy crisis, that oil is indeed running out and must be replaced by other energy sources, that most of the alternatives proposed will not work or not be economical, or may only work in the distant future. Energy presents many technical problems. Some of these are soluble, like that of nuclear waste disposal. But there are also problems insoluble by foreseeable technology, such as making large amounts of energy from solar heat in an economic manner. The public should not demand that technology solve an insoluble problem'. A contrary view was provided by Hannes Alfven, like Bethe a Nobel Prize physicist. The participants were told that he was also a nuclear physicist, but he is more accurately described as a magnetohydrodynamicist and so his words are much less weighty that those of Bethe.

There is often a tendency to be content with providing a forum for the expression of differing views and a reluctance to grasp nettles and formulate a clear response. Thus in the same meeting, Bishop Hapgood concluded that 'our group cannot put forward categorical recommendations. It would not feel justified in entirely rejecting, nor in wholeheartedly recommending large-scale use of nuclear energy'. He comments: 'One could scarcely be more judicious than that!' One could also say that it is weak and inconclusive. Was there no one strong enough to master the whole debate, separate the truth from the rubbish, and formulate a clear and convincing response to one of the urgent issues of our times?

Throughout the nuclear debate there is a reluctance to express results numerically. It is often not easy to do this, but approximate numbers, combined with some knowledge of their range of uncertainty, are infinitely better than no numbers at all. The debate can go on for ever if confined to the verbal level, and it can be resolved only at the scientific level. For example, in the document 'Energy for the Future', Karl Morgan discusses radiation exposure, drawing attention to the need to control medical doses, remarking that 'were we to completely eliminate the nuclear power industry to the end of the century, this would reduce the population dose less than would a 1% reduction in our average dose from medical applications. Many other examples are given in this chapter. Instead of decisions reached by experts on the basis of statistical analyses, many of the recommendations at WCC meetings are reached by voting. This may be democratic, but it may well be asked whether all those who voted on these vital matters were thoroughly conversant with the relevant scientific and technical arguments. As an example, the participants at the MIT Conference on 'Faith and Science in an Unjust World' voted in favour of a five-year moratorium on the construction of nuclear power stations. This proposal was endorsed by the Central Committee of the WCC, and yet when this question was raised by Janos Pasztor of the WCC during the study week of the Pontifical Academy of Sciences, it was emphatically rejected by the experts present.

In the same publication, P.J. Dyne considers radioactive waste management and says: 'I have examined my conscience long and hard over these matters and I can simply state my conviction: radioactive waste management can be done: it is a reasonable technical objective. If properly done the risks are minimal'.

Sigvard Eklund, the Director-General of the International Atomic Energy Agency, explains why we need nuclear power over the next three decades and probably as our main source of energy for the centuries. He remarks: 'I am very much disturbed by the growing tendency to mistrust the advice given by outstanding specialists. Instead, one often finds media, the public and even governments preferring to listen to self-appointed 'experts' from other fields or, worse, from no fields at all. The real experts are disqualified by their own expertise, whereas the non-experts are to be believed, because they are considered to be objective'.

Considered as a whole, these publications of the WCC provide a most valuable compendium of information, arguments and views on the desirability of nuclear power. They contain the texts of many lectures by real experts that repay careful study. Particularly laudable is the emphasis on the needs of the poor. On the whole, however, it must be said that there is a tendency to present the arguments and counter arguments without coming to definite conclusions and recommending effective actions.

There are many other statements by the Churches on the desirability of nuclear power, but it would be unduly repetitious to summarise and comment on them all. It may be useful, however, to list a few of them (Detailed analyses and summaries are given in 'The Churches and Nuclear Power', August 1984):

1. European Churches and the Energy Issue. Official Statements, Reports, Comments 1975–1979 (ed.), Friedholm Solms. Forschungssatte der Evangelischen Studiengemeinschaft, Heidelberg 1980. Included in this volume are statements from Protestant, Catholic, Evangelical, Reformed, Waldensian and Methodist Churches as well as various Synod, Councils of Churches and Ecumenical Councils.
2. Statement of the Committee of the US Catholic bishops. Published in Origins, NC Documentary Service, 23 April 1981, vol. 10, no. 45, p. 706. Summary and Critique in The Month, November 1982, p. 382.
3. The Christian Dimensions of Energy Problems. Papers delivered at a Conference held at Brunel university in April 1982. Commission for International Justice and Peace, 1983, p. 82.
4. Nuclear Energy: What are the choices? A Report by the Quaker Nuclear Energy Group. Published for the yearly meeting of the Religious Society of Friends (Quakers) by Quaker Home Service, London, p. 22.

10.4. General Assessment of Church Statements on Nuclear Issues

To be worthwhile, a Church statement on any matter should be accurate, balanced, well-informed and directed to a matter of public concern in a way that commands attention and respect, makes a specifically Christian contribution, corrects existing misunderstandings and makes specific recommendations for the direction of future policies.

By this standard the Church statements mentioned are, with a few exceptions, deplorably inadequate. Obviously they contain incidental remarks of value, but taken as a whole they are little more than an uncritical echo of the anti-nuclear campaign, embellished with a veneer of religiosity. No one with any knowledge of the real situation would take them seriously for an instant. They serve only to bring discredit on their authors and the Churches that issue them.

The exceptions deserve careful attention because they can help us to understand what is needed to make a useful contribution to the nuclear debate. The most notable are the Methodist Report Shaping Tomorrow and the Proceedings of the Study Week organised by the Pontifical Academy of Sciences.

The first and most important point is that both are based on a thorough and detailed scientific study by a number of experts. The sheer scale of the problem to be studied had been correctly assessed. It is simply not possible to say anything sensible on the strength of the opinions of a handful of people, often lacking any expert knowledge. The Methodist report was the result of the studies of some sixty scientists, technologists and engineers working for two and a half years. The Vatican study brought together about thirty experts of international standing. Such groups of experts can provide the range of knowledge and of different points of view to form the basis of an accurate and balanced statement.

Again and again one finds well-meaning people who completely underestimate the work needed to form an accurate and balanced judgement. They apparently think that on the basis of knowledge gained from a few newspaper articles or television programmes they can make public statements on the moral aspects of complex technological problems. Inevitably the result is a disaster.

It is essential to reach conclusions about the best course of action, and then ensure that they are widely distributed so that they can form the basis of action and further debate. The Methodist Report Shaping Tomorrow is well-written and attractively printed and should have a wide influence. Unfortunately this is likely to be largely cancelled out by the disgraceful Future Conditional which is also attractively printed and is presented in a way that will engage the interest of younger people. The Vatican report is a weighty tome that has received very little distribution and is almost entirely unknown. It thus seems that in the very few cases when reliable studies have been made they have not been recognised as such and given the publicity they deserve.

This is indeed a situation that occurs quite frequently. We recognise a need for more information and say that the subject should be studied. Then subsequently we find that the work has already been done, that the information is available in a publication that has been ignored. As a result, much time is spent re-inventing the wheel and so no progress is made. It should therefore be possible to make a valuable contribution simply by identifying the best publications and seeing that they are widely known.

It is rather easy to reconstruct the way statements on nuclear power, and indeed on many other subjects, come to be issued. The subject was recognised as one of importance, with moral implications. In many cases it is likely that Church authorities were urged to make a statement by political activists. The Church authorities then realised that difficult scientific and technical matters were involved, and so set up a committee, or perhaps took the advice of the nearest activist. Without realising the magnitude of the task, or the inadequacy of the means they had assembled to tackle it, they then proceeded to draft and issue a statement, convinced that they are making a serious and valuable contribution to the public debate. When it is politely pointed out to them that this is not the case, they are pained that anyone should question their abilities, brush aside the detailed comments made, throw a veil of secrecy over the whole subject, and lapse into impenetrable silence.

It is not difficult to predict what happens next. Scientists who were only too willing to put their knowledge at the service of the Church, but who find that they are ignored, decide that it is a waste of their time to do anything constructive and retreat to their professional work. (During a discussion with a friend, I asked whether he communicated his views on the topic we were discussing to the Church authorities. He replied that it was no use, as they never listen). It is nevertheless irritating to see rubbish published in high places, and many scientists see it as their duty to correct it. Denied a constructive role, they have nothing left but silence or destructive criticism. Since they occupy the high ground this can be done very effectively, and inevitably strengthens the unpopularity of scientists in clerical circles. Speaking in a more general context, the same point was made by Dr. John Mahoney when he said: 'There is an absolute and urgent necessity to raise the level of theological awareness, and not let the high ground be permanently occupied by embittered snipers.' (The Tablet, 12 May 1984, p. 452).

10.5. Misleading and Irresponsible Statements

Many statements on technological matters are subtly misleading or irresponsible, and it may be useful to give a few examples. Language is a delicate instrument, and it is very easy to make a statement that conveys a meaning different from what it actually says. The examples given below are taken from statements made by Church and other bodies on nuclear power, but similar ones may be found in other contexts. They include a whole range of statements from those that are just false but plausible to true statements presented in a way that conveys a false impression. Much depends on the context of a statement so, for instance, a cautionary remark in the midst of a balanced discussion is entirely acceptable, but taken out of context may be quite misleading. The falsity of many statements often appears only when numerical data are given, and the difficulty is that many people are unable to understand statistical statements.

(a) 'Our energy problems can be solved by more research on the renewable sources such as windmills'. This is wishful thinking. Many studies have shown that windmills are more expensive and dangerous than other sources, and they are also unreliable and environmentally offensive. Their efficiency can be somewhat improved by research, but not their unreliability. No amount of research can alter this.

(b) 'There should be a moratorium on nuclear power until it is made perfectly safe.' No source of power is perfectly safe, so all we can do is to make each source as safe as reasonably possible, and then choose the safest. Studies have shown that nuclear power is among the safest ways to produce energy. To do nothing is itself a decision whose hazards can be estimated. Thus the anti-nuclear campaign has already killed several hundred people by delaying the building of nuclear power stations which would reduce the number of more dangerous coal power stations.

(c) 'Nuclear reactors are emitting poisonous radiation that is shortening all our lives.' This is very probably true. A recent estimate, making the worst assumptions, is that statistically they shorten our lives by an average of about one second. For comparison, smoking one packet of cigarettes a day shortens life by about 3000 days.

(d) 'The purpose of this study is to list the arguments both for and against, and to leave it to the reader to reach a conclusion for future action.' This is considered to be a wise and judicious stance, but it is simply an abnegation of responsibility. If the committee of experts making the statement is unable to reach a decision, is it likely that non-experts will be able to do so?

(e) 'The most democratic way to reach a decision is by taking a vote.' This was once done after some lectures by experts. Once again, can we expect the views of non-experts to be better than those of experts?

(f) 'We can obtain all the uranium we need from seawater.' There is certainly a huge amount of uranium in the sea, but it is necessary to know how much it costs to extract and whether there are any environmental effects. The present indications are that it is too costly and produces large amounts of waste.

(g) 'A nuclear reactor contains about a thousand times the radioactivity released by the Hiroshima bomb.' This could well be true, but the essential point is not the amount of poisonous material but where it is, how it is contained, and is it likely to reach humans. The radioactivity released by the Hiroshima bomb went into the atmosphere, while that in a reactor is kept safely inside.

(h) 'Nuclear reactors are liable to explode like a bomb'. This is false. If the numbers of reactions increases the reactor expands and the reaction rate falls again. If however the reactor is badly designed, and is operated as irresponsibly as that at Chernobyl, when the safety devices were switched off, then the reaction rate can rise very rapidly, setting it on fire and releasing radioactivity into the atmosphere. To oppose nuclear power because this happened once is like forbidding the construction of ocean liners because of the Titanic disaster.

(i) 'Fusion reactors will be clean.' Certainly they will not produce any fission fragments, but the danger comes from radioactive materials, and these are produced in substantial quantities in fusion reactors.

(j) 'Nuclear power should only be developed with great caution.' This obviously true statement insinuates that nuclear power is especially dangerous. No one in their senses would suggest that anything should be operated carelessly.

(k) 'The public must be told about nuclear power'. Of course the public must be told, but this statement insinuates that it is being kept in the dark. The truth is that the public has been told a thousand times but does not understand because any accurate and responsible statements are drowned by propaganda and errors. It is a naïve and dangerous error to suppose that all that is necessary is to tell people the truth, and all will be well.

(1)   'The carbon dioxide emissions from wind turbines are less than those from nuclear power stations'. This may or may not be true, but it is irrelevant to the choice between them. The results from several studies are given in Table 6.2, and it is apparent that they differ considerably. However, the important point is that they are about a hundred times less than the emissions from fossil fuel power stations, and so the first objective should be to reduce dependence on fossil fuels, and then decide between nuclear and the other possibilities using the other criteria already discussed. It may also be mentioned that coal power stations emit more radioactivity than nuclear power stations, but in both cases the amounts are minuscule and so this is not a valid argument in favour of nuclear. As in so many other cases, it is necessary to look at the relevant numbers in order to reach a correct conclusion.

It is depressing that so many of those who take part in what is euphemistically called the nuclear debate repeat statements such as the above without any attempt to find out whether they are true or not. As I have frequently experienced, they are impervious to reasoned argument and refuse even to listen. They are so sure that nuclear is bad that they will not consider any arguments or take part in a real discussion. They propagate their views so assiduously that many people follow them and thus they exert a strong influence on Government policy.

There is nothing new in this, as Jenner (Miller 1983) discovered when he was faced with the cost in lives and suffering resulting from the activities of the anti-vaccinationists: 'One would think the statement of Facts, as they now stand before the Public from every Quarter of the Globe would blow away such stuff as these abominable people produce, like Chaff, but it is not so, or the Bills of Mortality would not exhibit weekly such horrid scenes of devastation from the Smallpox'. It would be understandable if such a man were to say: 'I have given you the means to eradicate smallpox; take it or leave it. If you prefer to remain in ignorance, to see your children die in torments, then let it be on your head and not on mine'. If he goes on fighting the ignorant mobs it is because he has in mind the reflection finely put by Donald MacKay (1979): 'The accusation of which we would have a right to be afraid on the Day of Judgement is this: "You knew what could be done in this way or that, and you did nothing"'. But if scientists have the duty to speak, does not the public have the duty to listen?

There are three basic errors underlying most of the statements by the Churches, namely innumeracy, unbalance and the failure to recognise objective truth. So many vital issues can only be settled by numerical estimates of cost, or safety, for example. Relying on rhetoric, emotion and even common sense is a recipe for disaster. Secondly, a statement must be balanced if it is to have any value. Thus if you are talking about the safety of power generation it is gravely misleading to consider only one source, pile up the arguments against it, and then jump to a conclusion. To reach a sound conclusion it is necessary to list all sources and compare them as objectively as possible. Thirdly, moral questions are not matters of opinion that can be settled by votes or rhetoric; they concern objective truth. It may not be easy to determine, it may be necessary to recognise the possibility of imprecision or error, but nevertheless it is always the final objective.

The public discussion of nuclear power began when the nuclear physicists who had worked on the atomic bomb during the war considered it their responsibility to do all in their power to inform the public of the dangers of nuclear weapons and the potentialities of nuclear power. Initially the response was very favourable and many countries started to build nuclear power stations, which proved on the whole very successful. It soon became clear that they were capable of providing the bulk of the power that the world so desperately needs. At this point strong political forces, for the reasons already mentioned, launched a strong and well-organised campaign against nuclear power. This was greatly strengthened by the accident at Three Mile Island and the disaster of Chernobyl. Several countries then resolved to stop building nuclear reactors and to phase out nuclear power as soon as possible. Worldwide the construction of nuclear power stations virtually ceased. In the same period people became increasingly concerned about the threats to the environment due mainly to the pollution from coal, oil and gas stations and other industrial processes. Acid rain and the possibility of climate change intensified these concerns. At first these were not taken very seriously, but now there is definite evidence from the melting of the Arctic to the increasing violence of tropical tornadoes that we are affecting the world climate. Governments are still reluctant to face this challenge and take effective action. The effects of the anti-nuclear campaigns are still strong, and Governments do not wish to court unpopularity by sanctioning the resumption of a nuclear power programme. Instead, they try to convince people that they can solve the world energy problems by relying on the so-called renewable energy sources, principally wind and solar, which are manifestly incapable of providing the power we need at an acceptable price. Unless realistic decisions are taken soon the situation will continue to deteriorate until more and more power cuts and natural disasters finally force Governments to act. The remedial actions inevitably take years to come into operation, and by then it may be too late.

References

Blanc-Lapierre, Andre (ed.) (1981) Energy and mankind, *Proceedings of A Study Week organised* (Pontifical Academy of Sciences), Vatican City, 10–15 November 1980. Academiae Scientiarum Scripta Varia No. 46.

Boyes, Edgar (ed.) (1981) *Shaping Tomorrow* (London: Home Mission Division of the Methodist Church).

Brungs, R., and Postiglione, M. (eds.) (2004) *Globalization: The Christian Challenge* (St. Louis: Itest Faith/Science Press).

Fernando, Kenneth (1981) *Energy for My Neighbour: Perspectives from Asia* (ed. Janos Pasztor) (World Council of Churches).

Gratzer, Walter (2002) *Eureka and Euphorias* (Oxford: Oxford University Press).

Hodgson, P.E. (1983) Nuclear power: Rome speaks, *The Clergy Review.* February, vol. LXXVII, p. 49.

Jenner, Brian (ed.) (1983) *Future Conditional: Science, Technology and Society* (London: Home Mission Division of the Methodist Church).

MacKay, Donald MacCrimmon (1979) *Human Science and Human Dignity* (London: Hodder and Stoughton, p. 61).

Miller, G. (ed.) (1983) Jenner: Letters (Baltimore: Johns Hopkins University Press).

Montefiore, Hugh (2004) Why the planet needs nuclear energy, *The Tablet,* **23** (October), p. 4.

Montefiore, Hugh and Gosling, David (1977) *Nuclear Crisis* (Prism Press).

Nishiwaki, Y. (1954) Bikini ash, *Atomic Scientists' Journal,* **4**, p. 9.

Oliver, John (2004) The Belmont Abbey Conference, London: The Newman Association.

Rotblat, Joseph (1955) The hydrogen-uranium bomb, *Atomic Scientists' Journal*, **4**, p. 224.

Sakharov, A. (1990) *Memoir* (New York: Vintage Press).

Weinberg, Alvin (1976) Facing up to Nuclear Power.

Appendix 1

Energy Units

The basic energy unit is the erg, defined as the work done by a force of 1 dyne moving a distance of 1 cm. The dyne is the force which, acting on a mass of 1 gram produce an acceleration of 1 cm per sec. A joule (J) is 10 ergs, and is also defined as the kinetic energy of a mass of 1 kg moving at 1 metre per second. Since this is very small for practical purposes, large multiples of the joule are frequently used, particularly the megajoule (MJ) (10(6)J), the gigajoule (GJ) (10(9)J), the pentajoule (10(15)J) and the exajoule (EJ) (10 (18) J). In the oil industry, the unit is the tonne of oil equivalent (TOE). A tonne is 1000 kg. 1 EJ = 22.7 TOE, or 1 TOE = 44 GJ. Also, 7.3 barrels = 12 tonnes of oil. One barrel of oil per day is 50 TOE per year. Rates of heat production are measured in watts. A watt is the rate of working of one joule per second, so the watt has 'per second' built into it. A kilowatt (kW) is 1000 watts, a megawatt (MW) is 10 (6) watts, a gigawatt (GW) is 10 (9) watts and a terawatt (TW) is 10 (12) watts. One kilowatt hour (kWh) is 3.6 MJ. 1 EJ per year is 32.2 GJ per second or 32.2 gigawatts. An energy unit used in nuclear physics is the electron volt (eV). One eV is 1.6×10 (–19) J. A million electron volts (1 MeV) is 1.6×10 (13) J. Each fission of a uranium nucleus releases 200 MeV = 3.2×10 (–13) J. One gram of uranium undergoing fission releases 82,000 MJ.

In discussion of energy the terms 'energy' and 'power' are often used indiscriminately. It is important to note that energy is power multiplied by time. Thus for example if an electric light bulb has a power of 100 watts and it is switched on for one hour it delivers 100Wh of energy.

Appendix 2

Nuclear Radiation Units

There are many different ways of measuring the intensity of nuclear radiation. Some of the units mentioned below are now obsolete, but they are included because they may still be found in earlier publications.

The roentgen is a measure of the ionisation produced in a tissue, and is such that one roentgen produces two billion ion pairs in a cubic centimetre of standard air. The number of atoms in a cubic centimetre of air is so large that a roentgen ionises only about one atom in ten billion. This unit was originally defined for X-rays and gamma rays, and later a similar unit, the rad, was defined for any ionising radiation. The rad corresponds to the absorption of a hundred ergs of energy per gram. Since a roentgen delivers about 84 ergs per gram the two units are very roughly the same. More recently, a new unit, the gray, has been introduced. This is defined as the radiation corresponding to an absorption of 1 joule per kilogram. Thus 1 gray is equivalent to 100 rad.

Another important unit is the curie, defined as the radioactivity of 1 gram of radium. This may be extended to other radioactive substances by defining the curie as the radioactivity of an amount of that substance that has the same number of distintegrations per second, thirty-seven billion, as a gram of radium. For must purposes this is an inconveniently large unit, so the millicurie and the microcurie, respectively one thousandth and one millionth of a curie, are often used instead.

Since some types of radiation cause more damage than others, a unit has been defined to provide a standard of comparison between them. This is the relative biological effectiveness (RBE), defined as the dose from 220 KeV X-rays causing a specific effect divided by the does from the radiation causing the same effect. To make it possible to define the relative effects of various radiations on man an effective quality factor Q may be used. This has the value unity for X-rays, gamma rays and beta rays, ten for neutrons and protons and twenty for alpha particles and other multiply-charged particles.

When considering the effects of nuclear radiations on man, it is also necessary to include the different sensitivities of the different organs of the body. This is done by defining a unit, the rem, that is the product of the absorbed dose in rads, the effective quality factor and any other modifying factor. The rem can also be defined as the dose given by gamma radiation that transfers 100 ergs of energy to each gram of biological tissue; for any other type of radiation it is the amount that does the same amount of biological damage. A more recent unit, the sievert, is defined as 100 rem. A millisievert is thus 0.1 rem.

Index